This book is to be returned on or before
the last date stamped below.

-4. NOV. 1992

20 MAR 1995
28 NOV 1995
1 NOV 1996
13 JUL 2000
30 MAR 2001

29 NOV
10 MAR 1989 1988

26 OCT 2005

3 0 MAY 1989   1 9 OCT 2006

-8

-8

26 APR 1993

13. JUL 1993,
29 APR 1994
17. FEB 1995

Lockhart, R.D.

By R. D. Lockhart
G. F. Hamilton and F. W. Fyfe

★

ANATOMY OF THE HUMAN BODY

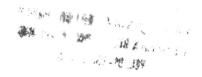

LIVING ANATOMY

A Photographic Atlas of
Muscles in Action
and Surface Contours

by

R. D. LOCKHART, Ch.M., M.D., LL.D., F.R.S.E., F.S.A.Scot.

*Emeritus Professor of Anatomy, Honorary Curator
Anthropological Museum, Aberdeen University,
formerly Professor of Anatomy, Birmingham University*

ff

faber and faber
LONDON · BOSTON

First published in 1948
by Faber and Faber Limited
3 Queen Square, London WC1N 3AU
Sixth edition 1963
First published in Faber Paper Covered Edition 1970
Reprinted 1971
Seventh Edition 1974
Reprinted 1979 and 1986
Printed in Great Britain by
Redwood Burn Ltd, Trowbridge, Wiltshire
All rights reserved

ISBN 0 571 09177 6

PREFACE

THE purpose of this book is to awaken the student's interest in studying, literally at first hand, muscles in action in the living body, indeed to divert him in some measure from textual descriptions in the belief that what he sees or feels demonstrated in the living subject under his own hand will have an instant appeal, carry conviction and impart confidence in his own ability to determine the muscles involved in particular movements. This is no new proposition. Two hundred years ago, the admirable Winslow, to use the title conferred by his countryman Duchenne, pointed to the fallacious deductions from experiments made upon the muscles of the cadaver instead of the living subject; over one hundred years ago, Duchenne, himself, brilliantly insisted upon muscles being studied not by scalpel upon the dead but by 'electrisation' upon the living, while almost fifty years later Beevor in his Croonian lectures had still to plead for the time to come when the action of muscles would be taught on the living body. 'Keep your eye on the ball' is sound advice. Keep your eye on the body, especially the living body, is the first principle in anatomy. These excellent principles have been frequently reiterated by Anatomists despite the recurring criticisms of their alleged cadaveric bias, but the lesson has usually lacked adequate illustration. 'A little picture is worth a million words', says the Chinese proverb: it is hoped these pictures may play a little part in the quickening task of raising anatomy from the dead.

R. D. LOCKHART

UNIVERSITY OF ABERDEEN
December 1947

PREFACE TO THE SIXTH EDITION

SINCE the last edition twenty-six new illustrations have been added.

Figures 203–206 demonstrate the loss of the elasticity of the skin; 207–212 illustrate the remarkable change in the disposition of certain areas of skin, as well as its elasticity in the act of raising the arm; 213–217 depict changes in facial features from infancy to old age.

Figures 25, 26, 44–49, 97 show development of the shoulder and trunk muscles in the 'catcher' of a circus trapeze act.

R. D. LOCKHART

UNIVERSITY OF ABERDEEN
August 1962

PREFACE TO THE SEVENTH EDITION

IN the twelve additional illustrations surface musculature is displayed in front, back, and side views, pages 94, 95, 96, whereby the student may test his ability in relating and identifying with the living model. Photographs and radiographs of the foot under different conditions, page 92, contrast effects upon the longitudinal arch. A series of footprints, of the same person, from birth to adult, show the 'growth of a footprint' and also walking and standing prints, as well as those from an aboriginal woman who has never worn shoes, and from an athletic girl prone to prevailing fashions in footwear, page 93.

R. D. LOCKHART

UNIVERSITY OF ABERDEEN
June 1974

INTRODUCTION

THE physician examining a patient makes the muscles speak for themselves and their nerves by resisting his patient's movements so that the muscles act more powerfully. Clinically, this is termed 'reinforcing' the action of the muscles. Whenever a muscle can be palpated or the results of its action demonstrated upon the living body, the student of anatomy should never fail to use the clinician's method. Suppose for example he grasps the subject's deltoid muscle with one hand and with the other holds the subject's arm abducted to the horizontal, while the model is told first to adduct then to abduct against the operator's resistance, the alternate softening and hardening of the muscle will present the student with one of the most important lessons and principles he ever will receive in myology. This method of rapidly and alternately throwing a muscle's action in and out applies a good test where doubt arises about the part played by a muscle in a movement.

It is not the design of this book to deal with every muscle, or even to depict all the activities of an individual muscle, but merely to encourage the student in finding out more for himself.

The beginner has little difficulty with the classifications of muscular activity in the orthodox text. He readily appreciates that in flexing the elbow the flexors, biceps and brachialis, doing the work are the prime movers, while the extensor, triceps, able if necessary to prevent the movement, pays out as the antagonist; that in raising his lower limbs from the floor with the body supine, his abdominal recti muscles become very sore playing the most exhausting part, not directly in raising the limbs, but as fixation muscles in preventing the pelvis rotating forward under the leverage of the legs; that the secret of firmly clenched fingers is the synergistic action of the extensors of the wrist in keeping the wrist bent well back, that the trick of forcing an opponent to drop his clenched weapon is to flex his wrist acutely, thereby cancelling the synergistic action at the wrist, and in turn straining the extensors of the fingers beyond yielding point till they overcome the finger flexors when the fingers must fly open (p. 23); that all these four roles may be played in unison by different muscle groups in certain movements and interchanged between different groups in other movements; that muscles may be in antagonistic teams the one moment and in the same team the next, for example, the anterior and posterior tibials, antagonists for flexion and extension of the foot, yet co-operate for its inversion.

The average student has more difficulty in explaining what is meant by describing a muscle as a voluntary muscle, because, to tell the truth, the muscle is not voluntary at all. The description is merely a convenient convention applied to muscles that take part in movements made of our own free will, indeed the term voluntary applied to the muscle is a figure of speech where the quality of the movement becomes attributed to the muscle. With the upper limb raised to the horizontal in front of the body firstly let the forearm rise to a right angle and secondly lower to meet the arm. In the first part of the movement the flexors do the work pulling the forearm up, but in the second part the extensor, triceps, does the work letting the forearm down; we execute a simple continuous voluntary movement but we are not in the least conscious of this delicate and exquisitely timed neuromuscular co-ordination whereby one group of muscles is thrown into action and the other out as the plumb line is crossed. The most striking, if pathetic, illustration of this point occurs when a patient's forearm with elbow flexors paralysed, is raised to the vertical, and he smiles with delight, imagining his muscles have recovered, as he suddenly gains control of the continued flexion movement by his intact triceps. Another illustration is to press the fore-finger and thumb together very gently when it will be found that the other fingers of the same hand are not freely mobile, and that the firmer the fore-finger and thumb are pressed together the stiffer the other fingers become. This exercise also serves to show that even in simple movements, several muscles are used, and that probably nowhere in the body does a single muscle ever act alone. Again, when the eyes are closed tightly, the slight drumming in the ears is due to the associated action of the stapedius muscles upon the stirrup bones of the ears.

The majority of students do not fully realise the influence of gravity and posture upon muscular activity. When in the erect posture the subject holds an arm straight out in front, the posterior spinal muscles may be seen and felt contracting; not that they are directly concerned in raising the arm but because they must keep the trunk balanced erect against the leverage exerted by the advanced arm (cf. p. 32). This may be proved by the model, with the arm still outstretched, leaning slightly backwards when the spinal muscles will at once become flaccid. The sternomastoids flex the neck when we rise from the supine position (p. 13), but in the erect posture they never flex the neck unless there is resistance to the movement. This principle is well seen in the exercise already cited in the previous paragraph where, with the upper limb in the horizontal position, the forearm is first flexed to a right angle and secondly lowered to touch the upper arm. In the first phase the flexors work pulling the forearm up against gravity, but the moment the vertical line is passed, they go out of action as gravity begins to assist the movement, and the extensor, triceps, takes over the work paying out and lowering against gravity.

In several of the foregoing examples the point has been stressed that a muscle may be performing the real work of a movement not only when it is contracting but also when it is paying out. In texts upon physical training and physiotherapy very convenient terms have been used for many years describing these two working actions as the 'concentric' and 'excentric' actions respectively. These terms are not used in physiology and they are not descriptive or accurate; probably contraction and decontraction would have been more acceptable generally.[1] However, 'concentric' and 'excentric' are now time honoured, and save time but their full implications can be realised only from such cumbersome descriptions as the following account of the deltoid's activity in raising and lowering the arm. When the arm is being raised the deltoid can be felt firm—it is working hard and getting shorter or contracting. (Conversely the muscles antagonistic to the deltoid are not working, are soft and getting longer.) When the arm is being gradually lowered the deltoid can still be felt firm—it is still working hard in controlling the gravitational descent of the arm but getting longer as it pays out, like a crane lowering a weight. (Conversely the muscles antagonistic to the deltoid are not working, are soft and getting shorter.)

Here it is opportune to beware of mechanical calculations upon the skeleton and cadaver showing that a muscle is in the anatomical position to take part in a movement. This is no guarantee that the muscle belongs to the group executing this movement when the will desires it. Further, a muscle may be stimulated electrically to perform a movement in which it does not participate under the will, for example, when the arm is held erect the clavicular fibres of pectoralis major are in a position to depress the arm and under electrical stimulation they do so, but not during the voluntary movement of depressing the arm.

In the photographs presented in the following pages the same action is frequently shown from slightly different aspects and by different models. Accordingly it is unnecessary to place pointers and labels on all the plates. The student will test his knowledge upon the intact plates and note individual variations in physique and function. This fact of individual variation sometimes causes difficulty in determining the muscles involved in the same movements in different models. Certainly there is beauty, precision and economy in the trained movement of the skilled athlete, while the untrained person wastes the activity of many muscles unnecessarily in exaggerated efforts. Movements must be carried out several times and upon several models. Apart from different execution there is often different muscular endowment, for example many normal people, as is well known, cannot touch their toes while the knees are straight or perform a high kick, because of tight hamstring muscles, but the fact that tight pectoral muscles cause remarkable discrepancies between different persons in the degrees through which the arms may be raised is not always appreciated (cf. pp. 37, 61, 62). In some cases a striking contrast is obtained by carrying out opposite exercises upon the right and left sides of the body at

---

[1] In a personal letter Professor Adrian writes that he discussed the question with Sir Charles Sherrington who ingeniously suggested contraction and decontraction.

the same time, for example, raising one arm and lowering the other against resistance (pp. 11, 35) engages different parts of the pectoral muscles producing a lop-sided appearance, and also causes marked asymmetry of the thoracic cage.

Familiarity with the dissected muscle is necessary but not infallible in recognising a muscle in the living subject. The medial head of the triceps may be puzzling even with the dissected arm side by side with the living subject, while it is a peculiar fact that the great majority of students invariably delineate the lower border of trapezius by running their finger from the last thoracic spine straight to the tip of the acromion, instead of showing the muscle as a very narrow band in its lower part reaching to the medial end of the spine of the scapula.

As for technical points in the photography, it was found that individuals in a high state of physical development do not necessarily yield the best photographs for the study of muscles and contours. Leanness rather than development, and a certain quality of skin colour favour the delineation. Most of the photographs were taken by artificial light with an ordinary quarter plate camera, except those of the eye which were taken with the Leica copying arm device and the Summar F2 lens. In the use of artificial light it is surprisingly easy to obtain dramatic art photography effects that are useless for the study of anatomy. These illustrations are selected from more than 1400 negatives, and while some are the results of repeated effort for improvement, others are the more fortunate first attempts.

ACKNOWLEDGEMENT

To the student models, some sixty all told, who patiently assisted in the production of these pictures, thanks are especially due, for many had to suffer our initial trials in photography and experiment without even the consolation of selection.

Colleagues in the department took a personal interest, in the full sense of the term. Criticism, argument and demonstration followed hot upon each other's heels. I am most indebted to Professor Forest W. Fyfe for his continued eager interest and suggestions, and to Drs. William Watt and Charles Anderson for their cheery assistance from day to day.

The photography achieved by the department's senior technician, Mr. Alexander Cain, A.R.P.S., required enthusiasm as well as judgment, skill and patience, well tried in our repeated attempts for improvements. It is fitting to acknowledge that these efforts would have been to little purpose in the absence of the ability and pride the blockmaker, the printer and the publishers display in their work.

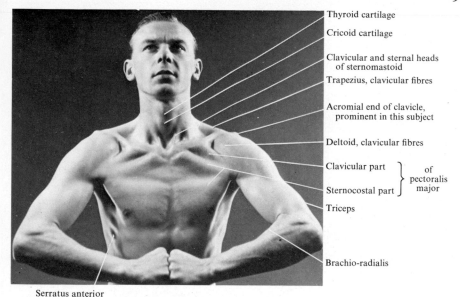

Thyroid cartilage

Cricoid cartilage

Clavicular and sternal heads
of sternomastoid

Trapezius, clavicular fibres

Acromial end of clavicle,
prominent in this subject

Deltoid, clavicular fibres

Clavicular part ⎫
⎬ of pectoralis major
Sternocostal part ⎭

Triceps

Brachio-radialis

Serratus anterior

FIG. 1. Pectoral muscles adducting against resistance
of hands

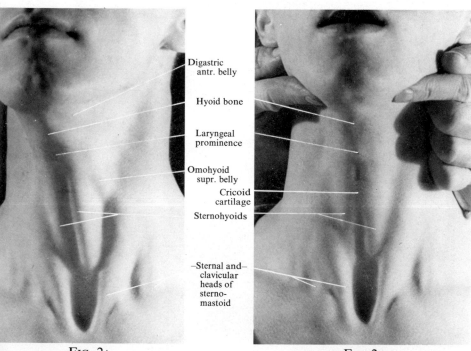

Digastric
antr. belly

Hyoid bone

Laryngeal
prominence

Omohyoid
supr. belly

Cricoid
cartilage

Sternohyoids

—Sternal and—
clavicular
heads of
sterno-
mastoid

FIG. 2A
Chin raised and muscles set

FIG. 2B
Chin lowered against resistance

10

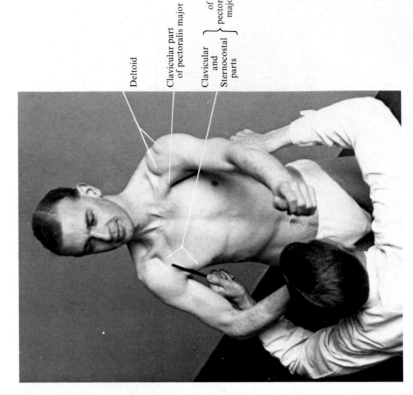

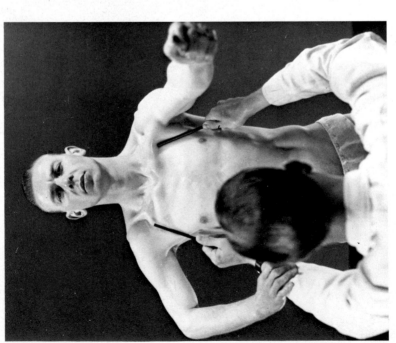

FIGS. 3 and 4. Actions of the pectoralis major muscles

The subject's right arm is being adducted against the operator's resistance, and the whole of the right pectoralis major muscle is active and tense, so that the pencil cannot be depressed into the muscle. The subject's left arm is being raised forwards (flexion at the shoulder), only the clavicular part of pectoralis major is active, and the pencil can easily be embedded in the inactive sternocostal part. The anterior fibres of deltoid are also working

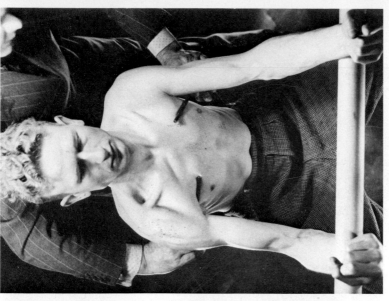

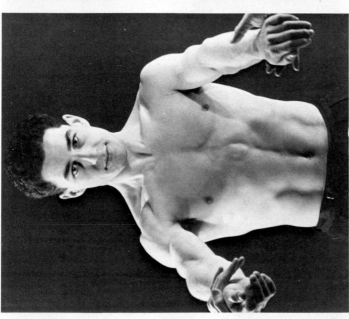

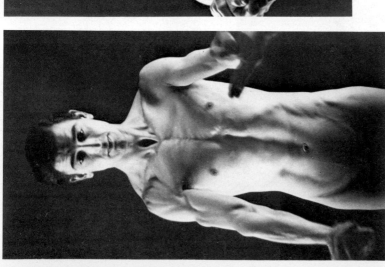

FIGS. 5, 6 and 7. Actions of the pectoralis major muscles

The subject's right arm is being raised forward against resistance (flexion at the shoulder). The subject's left arm is being depressed against resistance (extension at the shoulder) The clavicular part of pectoralis major is active in the right arm and inactive in the left arm. The sternocostal part is inactive in the right arm so that the pencil may be depressed into the muscle, and active in the left arm, so that the pencil cannot be impressed into the muscle. The upper fibres of trapezius, the anterior fibres of deltoid, and the biceps are active on the subject's right side, flaccid on his left. Apart from the lop-sided appearance due to the different parts of the pectorals engaged, there is also marked asymmetry of the thoracic cage in this exercise, Figs. 5 and 6

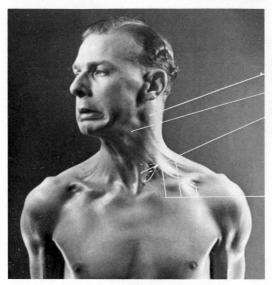

Sternomastoid

Platysma

Sternal and clavicular heads
of sternomastoid

Platysma

FIG. 8. Voluntary tension of platysma

Note its fibres streaming from the angle of the
jaw downwards across the sternomastoid, clavicle
and shoulder

The muscle may produce these features and facial expression in some competitors towards the end of a race, and it has been suggested that in strong inspiratory effort the taut muscle prevents retraction of the tissues at the root of the neck from compressing the veins and impeding the venous return. However, similar effects are seen in emotional distress, and moreover the muscle is sometimes absent

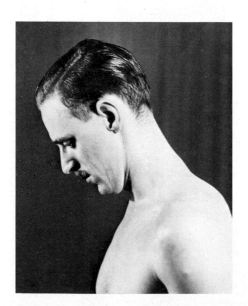

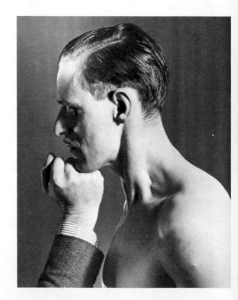

FIG. 9. Sternomastoid inactive in flexing head and neck

This movement is secured by the muscles in the back of
the neck relaxing under tension

FIG. 10. Sternomastoid active in flexing head and neck against resistance

In this case the muscles in the back of the neck are flaccid

Contrast with Figs. 11, 12, 13, 14, 21 and 22

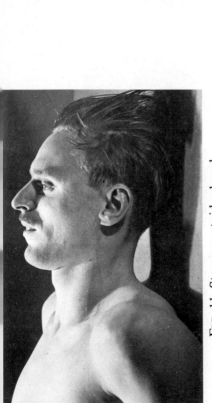

FIG. 11. Sternomastoid relaxed

FIG. 12. Sternomastoid flexing head and neck against gravity

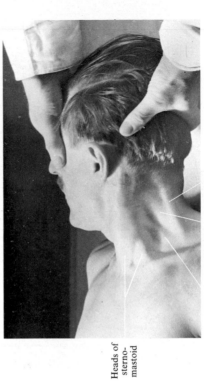

FIG. 13. Sternomastoid active in flexing head and neck against gravity and in rotating face to opposite side

Heads of sterno-mastoid

Scalenus medius Splenius Trapezius

FIG. 14. The action in Fig. 13 intensified by resistance

Contrast with Figs. 8, 9, 10, 21, 22

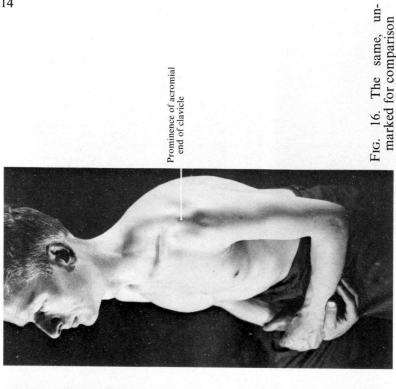

Prominence of acromial end of clavicle

Coracoid process
Tip of acromion
Angle of acromion
Deltoid

Fig. 15. Clavicle, coracoid process, acromion and spine of scapula outlined upon the skin

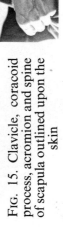

Fig. 16. The same, un-marked for comparison

CONTOURS OF THE SHOULDER REGION. SUBJECT AT REST

Note: It is a surprising fact that many senior students retain the erroneous impression that the clavicle, and not the acromion, comes to the tip of the shoulder region, and frequently they are unable to identify the position of the coracoid process. Compare with Figs. 17 and 18

FIG. 18. The same, unmarked for comparison

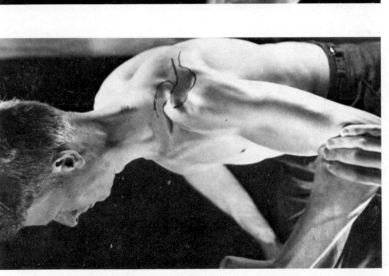

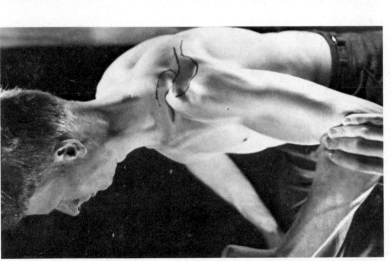

FIG. 17. Clavicle and acromion outlined on the skin

CONTOURS OF THE SHOULDER REGION. DELTOID ACTIVE

The acromion process is outlined by the fibres of the deltoid abducting against resistance: the lines of the tendinous septa in the muscle are also evident

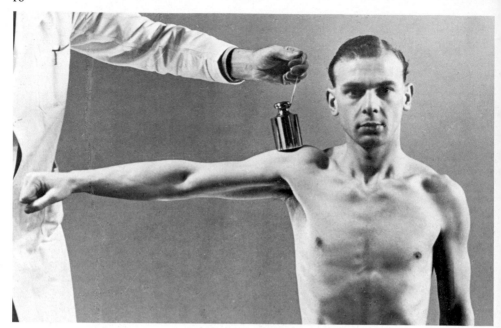

FIG. 19. Deltoid active

Whether the arm is being raised or lowered the weight rides upon a firm muscle, because the deltoid not only raises the arm but also controls its gravitional descent, respectively contracting or paying out under tension as a crane lowers a weight. The student may easily grasp this principle by raising and lowering one arm while he feels its firm deltoid with the opposite hand. Then he may demonstrate the contrast in Fig. 20 by trying to bring his arm down against an obstacle, when the deltoid becomes suddenly flaccid. In this connection see 'concentric' and 'excentric' action, introduction, page 7

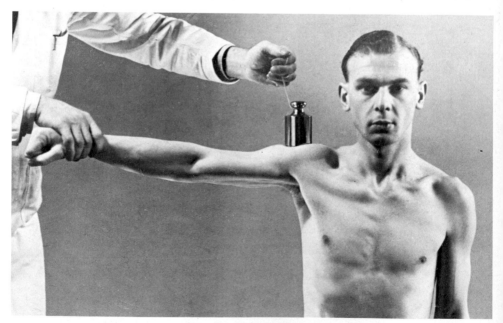

FIG. 20. Deltoid flaccid

The weight becomes embedded in the muscle when the antagonists such as the pectorals and latissimus dorsi pull down the arm against resistance. Contrast the active pectoral with its condition in Fig. 19

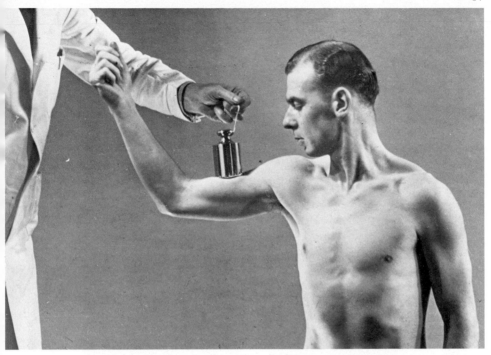

FIG. 21. Action of biceps as a supinator of the forearm

The weight rides upon the firm rounded muscle. Contrast with Fig. 22

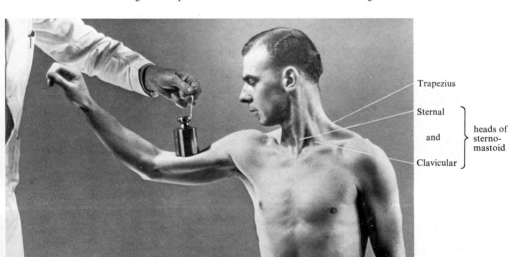

Trapezius

Sternal

and ⎫ heads of sterno-

Clavicular ⎬ mastoid

FIG. 22. In pronation of the forearm the weight tends to sink into a flattened biceps

Note the delineation of sternomastoid in turning the head in Figs. 21 and 22 and compare with Figs. 8–14

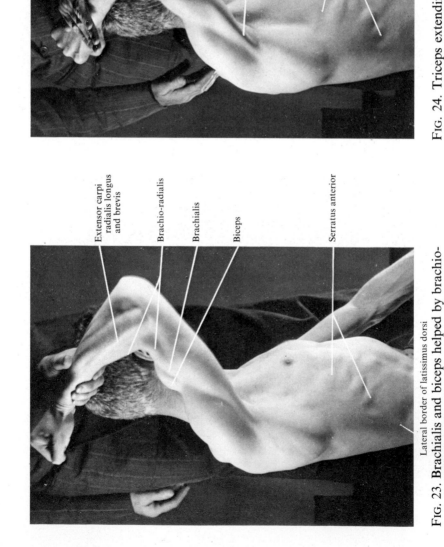

Triceps lateral head

Triceps long head

Serratus anterior

Fig. 24. Triceps extending the elbow against re-sistance, while biceps is flaccid

Compare with Fig. 23. Note the activity of serratus anterior in steadying the scapula

Lateral border of latissimus dorsi

Extensor carpi radialis longus and brevis

Brachio-radialis

Brachialis

Biceps

Serratus anterior

Fig. 23. Brachialis and biceps helped by brachio-radialis flexing the elbow against resistance

(Other muscles may take a slighter part in this action.) Triceps is flaccid but tends to bulge, sagging by gravity in this position of the arm.
Contrast with fig. 24

Note long and lateral heads of triceps and activity of serratus anterior, external oblique and latissimus dorsi muscles. See Fig. 41

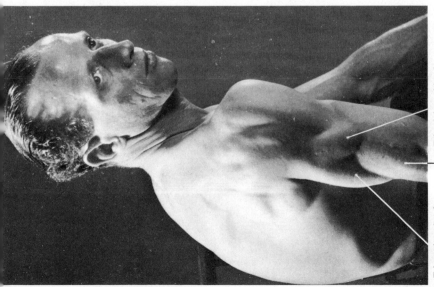

Triceps long head
Triceps common insertion tendon
Triceps lateral head

FIGS. 25 and 26. Good development of triceps, latissimus dorsi and serratus anterior muscles in the 'Catcher' of a circus flying trapeze act. The same model is shown in Figs. 44–49 and 97

20

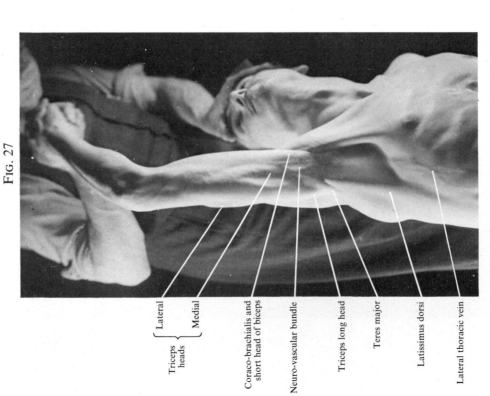

FIG. 28

Triceps heads { Lateral
Medial
Long

Teres major

Latissimus dorsi

FIG. 27

Triceps heads { Lateral
Medial

Coraco-brachialis and
short head of biceps

Neuro-vascular bundle

Triceps long head

Teres major

Latissimus dorsi

Lateral thoracic vein

Note biceps running into axilla posterior to pectoralis major, and
the position of the neuro-vascular bundle

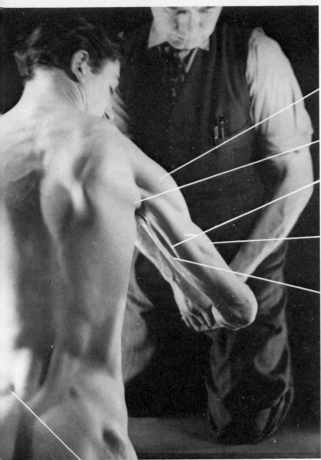

Deltoid

Teres major

Long ⎫
         ⎬ Triceps
Lateral ⎬ heads
         ⎭
Medial ⎭

FIG. 29. Triceps extending
elbow against resistance

Dimple indicates site of posterior
superior iliac spine

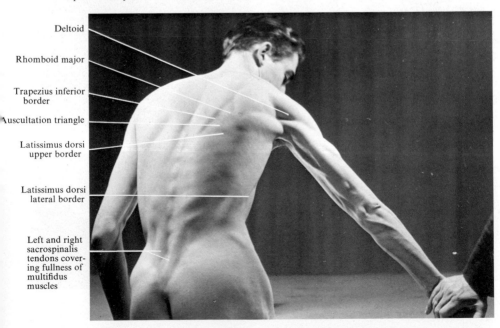

Deltoid

Rhomboid major

Trapezius inferior
border

Auscultation triangle

Latissimus dorsi
upper border

Latissimus dorsi
lateral border

Left and right
sacrospinalis
tendons cover-
ing fullness of
multifidus
muscles

FIG. 30. Extension of elbow and shoulder joints against resistance

Note the activity of the three heads of triceps, the posterior fibres of deltoid, teres major and latissimus
dorsi
Note: In the so-called triangle of auscultation the muscle boundaries separate slightly when the shoulders
are forward so that the chest wall becomes subcutaneous

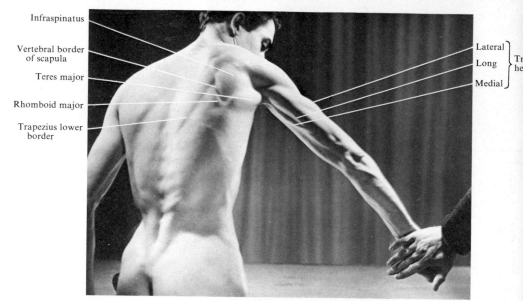

Infraspinatus

Vertebral border
of scapula

Teres major

Rhomboid major

Trapezius lower
border

Lateral
Long
Medial

FIG. 31. Triceps extending elbow against resistance

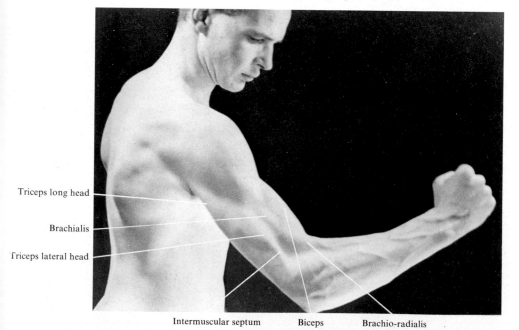

Triceps long head

Brachialis

Triceps lateral head

Intermuscular septum   Biceps   Brachio-radialis

FIG. 32. Muscles of arm 'set' to show intermuscular septum

23

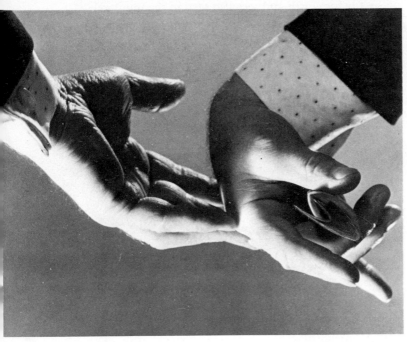

FIG. 33

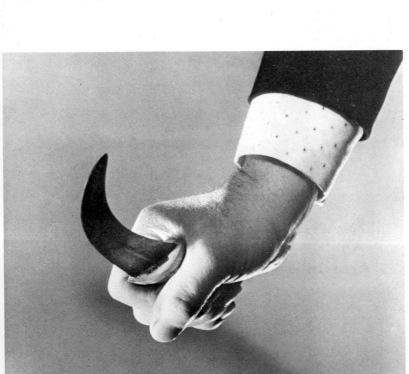

FIG. 34

MUSCULAR SYNERGY OR COMBINED ACTION

The firmly clenched fist is always bent backwards (extended) at the wrist, Fig. 33, but the fingers lose their grip and open when the wrist is forcibly bent forwards (flexed), a method of compelling an assailant to drop his weapon, Fig. 34. This is due to the extensor tendons being stretched to their limits. They cannot stretch enough to allow full flexion at the wrist with full flexion of the fingers, actions, which unopposed flexors would produce. Accordingly these two rival muscle groups co-ordinate in a precise synergic or combined action whereby the extensors control, or fix, the proximal wrist joint while the flexors clench the fingers at the distal joints

THE PERMANENT FLEXURE LINES OF THE PALM

(the romantic 'heart', 'life' and 'head' lines of the fortune teller which, however, are present before birth and are also seen in the monkey)

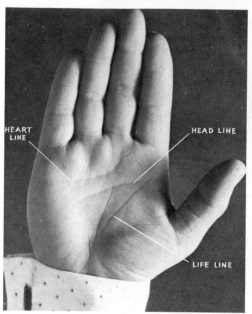

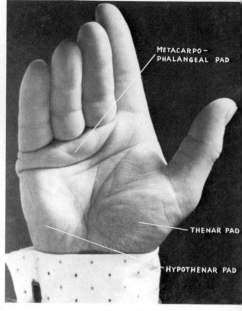

FIG. 35

FIG. 36

Along these palmar creases the skin is bound to the deep fascia by tough fibrous strands which quilt the fatty superficial fascia into pads, thereby affording the hand a variable yet glove-like grip upon diverse tools

The 'heart line' formed at the site of the metacarpophalangeal joints is accentuated by their movement and thereby swells the metacarpo-phalangeal pad. While in monkeys this line runs across the roots of all four fingers, in man it runs between the middle and forefingers, an indication of the singular mobility achieved by the human index finger

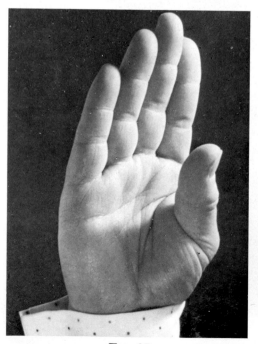

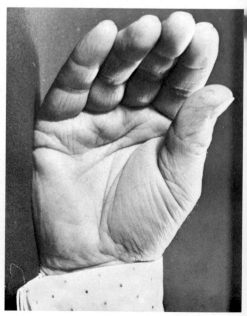

FIG. 37

FIG. 38

The 'lifeline' deepens with the movement of the thumb to the hand and the thenar eminence rises in relief

The 'head line' deepens as a compensating line between the other two

25

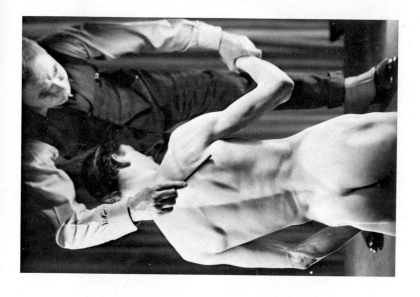

Fig. 40. Infraspinatus contracted (and resisting pencil) when subject's right arm is rotated laterally against resistance

Fig. 39. Infraspinatus flaccid (and pencil embedded) when subject's right arm is rotated medially

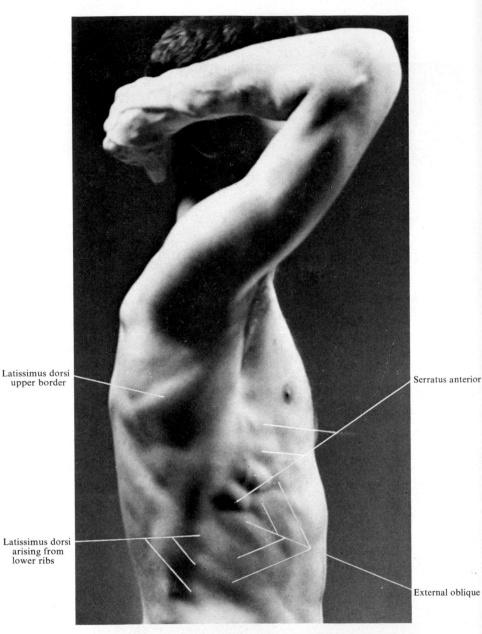

Latissimus dorsi
upper border

Serratus anterior

Latissimus dorsi
arising from
lower ribs

External oblique

FIG. 41. Serratus anterior and external oblique have been
'set', and latissimus dorsi pulls against resistance

Note the external oblique interdigitating with serratus anterior superiorly and
latissimus dorsi inferiorly, and also how the upper border of latissimus dorsi straps
down the scapula's inferior angle. Compare Figs. 42, 26

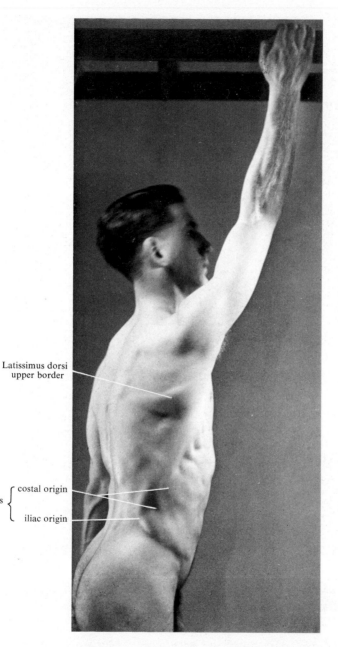

Latissimus dorsi
upper border

Latissimus
dorsi
{ costal origin
  iliac origin

FIG. 42. Latissimus dorsi pulling upon
the arm

28

Teres major

Vertebral border
of scapula

Triangle of auscul-
tation

Trapezius lower fibres

Upper and lateral
borders of
latissimus dorsi

FIG. 43. Latissimus dorsi pulling arm down against resistance

Compare Figs. 30, 41, 44-47, 50. Note how latissimus dorsi winds round the fullness of teres major in forming the posterior fold of the axilla

FIG. 47   FIG. 46 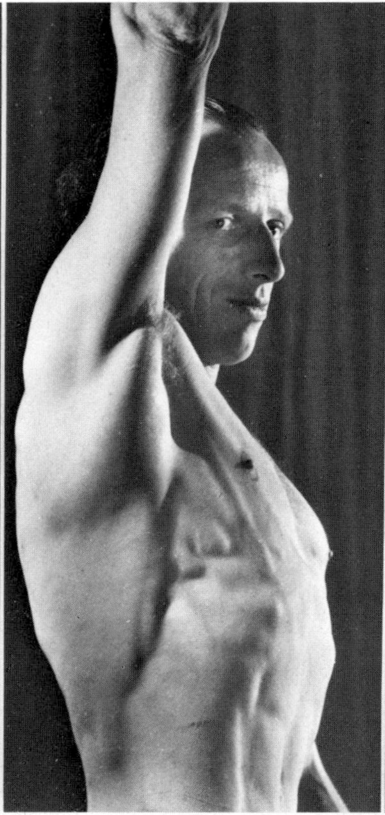   FIG. 45   FIG. 44

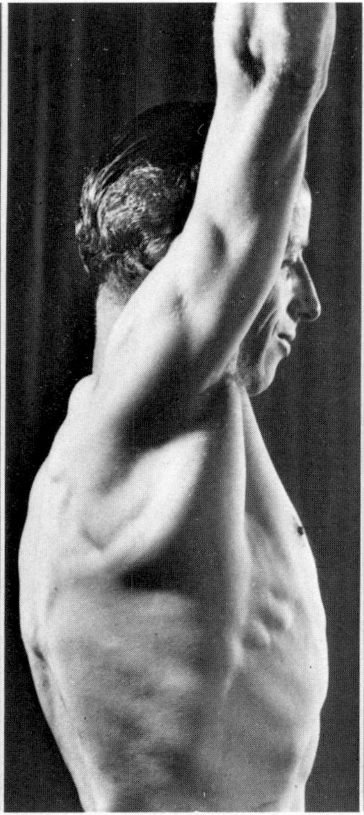

Four figures in sequence to show the latissimus dorsi muscle sweeping from the vertebral spines under trapezius, over the inferior angle of the scapula, Figs. 44, 45, and curving round teres major in forming the posterior fold of the axilla. Its lower and lateral fibres ascending vertically from tendinous attachment to the iliac crest are augmented by muscular slips from the lower four ribs interdigitating with external oblique Figs. 46, 47. This model, the 'Catcher' in a circus flying trapeze act, is seen also in Figs. 25, 26, 48, 49, 97

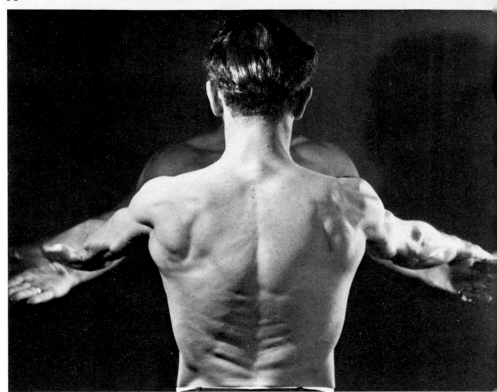

FIG. 48. Both arms lowered against resistance. Note activity of latissimus dorsi muscle as they pull on the vertebral spines through the posterior layer of the lumbar fascia

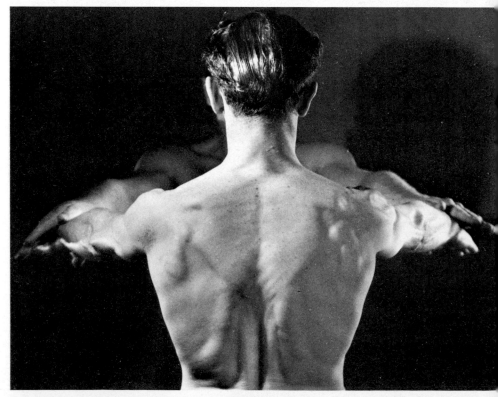

FIG. 49. Erectors of spine active against leverage exerted as both arms are raised against resistance

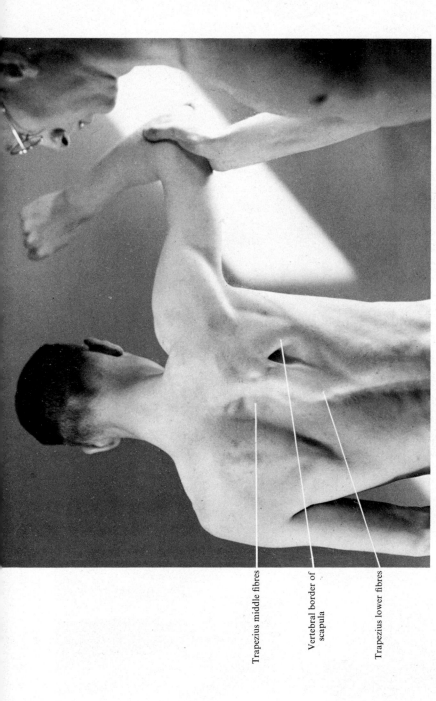

FIG. 50. The arm held in the horizontal position is being pushed up, and the subject resists by contracting the lower fibres of trapezius to pull the shoulder down

Note the marked depression over the triangle of auscultation. Contrast with Fig. 43

Trapezius middle fibres

Vertebral border of scapula

Trapezius lower fibres

FIG. 51   FIG. 52   FIG. 53

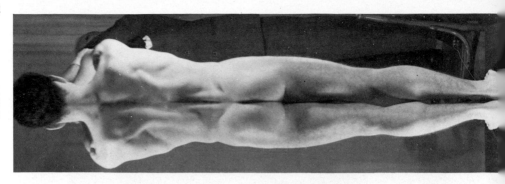

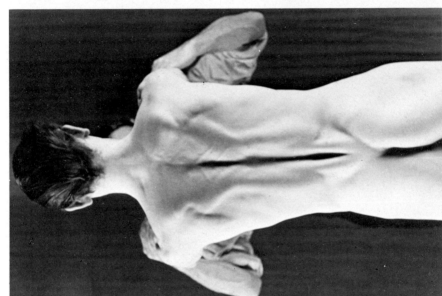

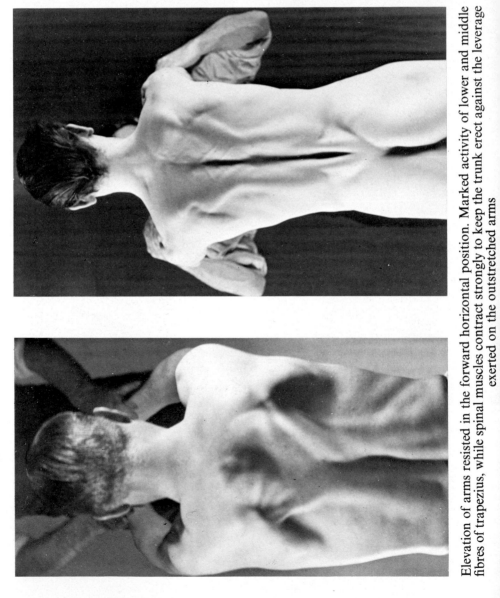

Elevation of arms resisted in the forward horizontal position. Marked activity of lower and middle fibres of trapezius, while spinal muscles contract strongly to keep the trunk erect against the leverage exerted on the outstretched arms

FIG. 53

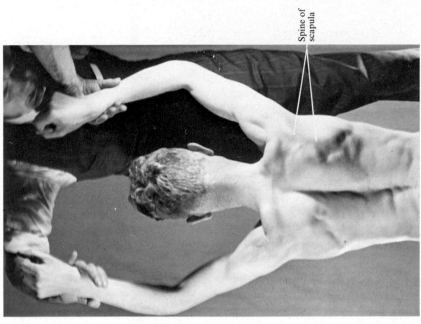

Spine of
scapula

FIG. 54

Elevation of the arms by abduction resisted. Activity of trapezius. Here the erect posture is not threatened
and the spinal muscles are not active as in Figs. 51, 52, 53

Note position of scapula and the fact that the spine of the scapula is practically in line with the shaft of the humerus as the arm becomes
erect. (Figs. 50, 51, 54 are the same person)

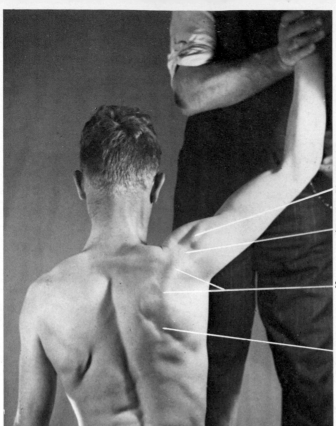

FIG. 56. Elevation of
arm resisted
Note position of scapula and
activity of trapezius and
deltoid

Deltoid

Acromion

Spine of scapula

Line of vertebral
border of
scapula

FIG. 57. The lower fibres of trapezius are
active, pulling down the scapulae, or rather,
preventing the scapulae being pulled up, as
the subject hangs his weight from the
operator's neck; the upper fibres are flaccid
to pressure from the operator's fingers

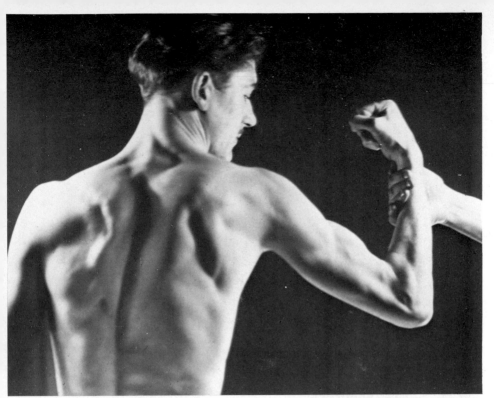

FIG. 58. Note the trapezius muscle fibres in relief against their diamond-shaped tendinous area between the shoulders

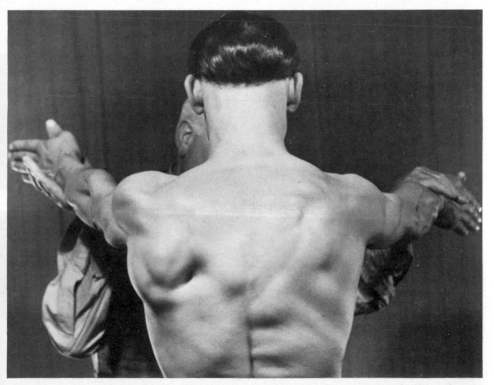

FIG. 59. Actions of latissimus dorsi and trapezius muscles contrasted. The left arm, pulled down against resistance, shows latissimus dorsi in relief curving round the inferior angle of the scapula to reach the axilla, but trapezius is passive. Contrast the right arm raised against resistance with trapezius firmly contracted, and latissimus dorsi passive

Mobility of scapulae with arms dependent

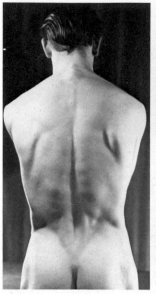 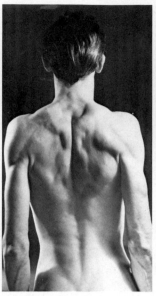 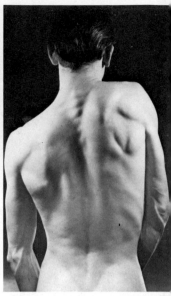

FIG. 60. Scapulae drawn forward by crossing arms in front

FIG. 61. Scapulae braced back by trapezius and underlying rhomboids

FIG. 62. Right scapula raised, left lowered

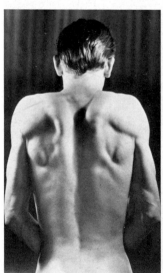 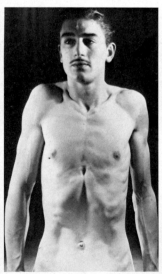 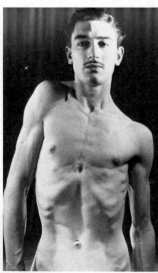

FIG. 63. Scapulae raised by upper fibres of trapezius, underlying rhomboid and levator scapulae muscles

FIG. 64. Front view of Fig. 63

FIG. 65. Front view of Fig. 62

Note range of clavicular movement

Mobility of scapulae with arms erect

Fig. 66. Arms erect at full stretch. Scapulae high

Fig. 67. Arms erect, controlled movement with scapulae low

Fig. 68. Right arm full stretch. Left arm controlled

Fig. 69. Front of Fig. 68

The arms may be raised vertically in a controlled movement in which the scapula is lower and the clavicle rises very little at its acromial end. Indeed, in some persons with tight pectoral muscles only the 'controlled' elevation is possible

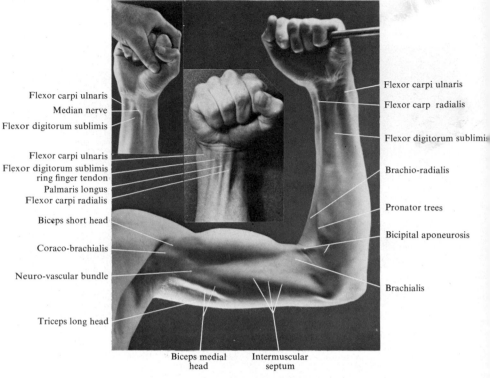

Flexor carpi ulnaris
Median nerve
Flexor digitorum sublimis

Flexor carpi ulnaris
Flexor digitorum sublimis
ring finger tendon
Palmaris longus
Flexor carpi radialis

Biceps short head

Coraco-brachialis

Neuro-vascular bundle

Triceps long head

Flexor carpi ulnaris
Flexor carp radialis

Flexor digitorum sublimi

Brachio-radialis

Pronator trees

Bicipital aponeurosis

Brachialis

Biceps medial
head

Intermuscular
septum

Flexors at wrist and elbow

In Fig. 70 the median nerve escaping from under flexor carpi radialis
to lie upon sublimis tendons is seen well because palmaris longus is
absent

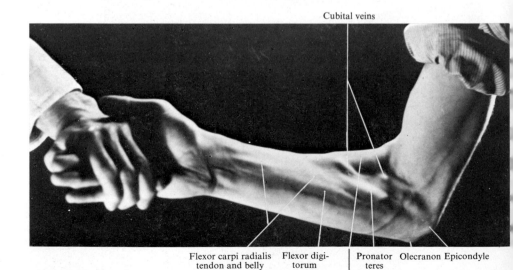

Cubital veins

Flexor carpi radialis
tendon and belly

Flexor digi-
torum
sublimus

Pronator
teres

Olecranon Epicondyle

Brachio-radialis

FIG. 73. Pronation attempted against resistance

FIG. 74. Flexion of meta-
carpo-phalangeal and interphalangeal joints of all the
fingers against resistance

The lumbrical muscle is not so active
as in Fig. 75; flexor tendon bundles taut
and visible

Flexor sublimis and
profundus tendon
bundles

First lumbrical

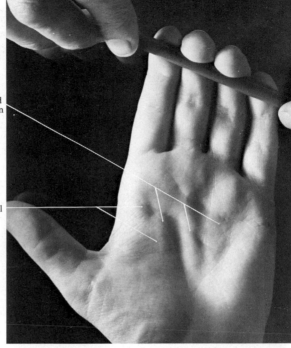

Flexor tendons
not seen

First lumbrical

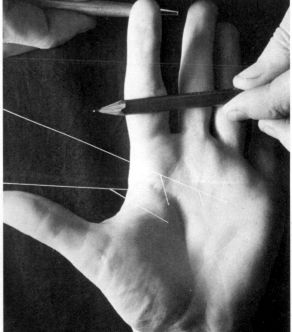

FIG. 75. The index finger
against resistance tries to
flex the metacarpo-phalangeal joint and extend the
interphalangeal joints (other
fingers inactive)

Lumbrical muscle active; flexor tendons
invisible. Contrast with Fig. 74

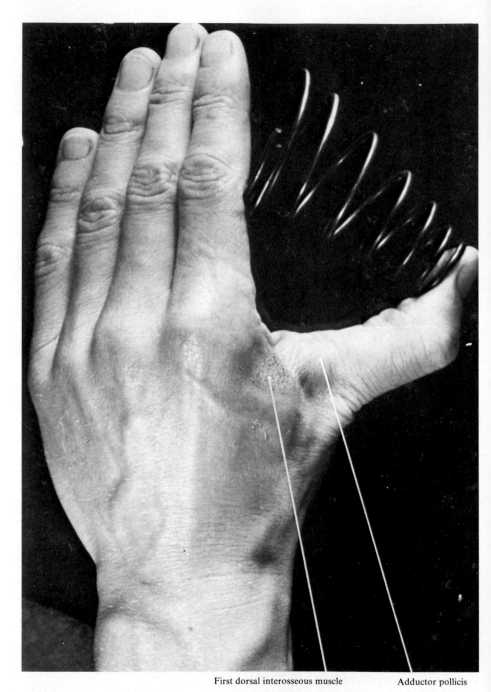

First dorsal interosseous muscle   Adductor pollicis

FIG. 76. Adduction of thumb and abduction of index finger against resistance (there is hyperextension of the metacarpo-phalangeal joint of the thumb)

First dorsal interosseous m. and
Head of Ulna | its tendon

Subcutaneous edge of
dorsal expansion of
extensor tendon

Flexor pollicis
brevis

Abductor pollicis
brevis

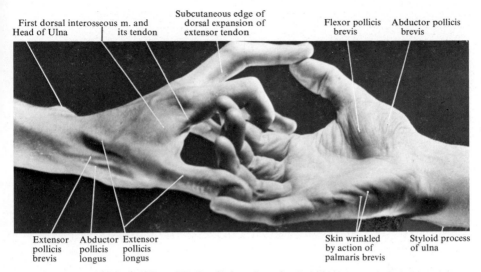

Extensor
pollicis
brevis

Abductor
pollicis
longus

Extensor
pollicis
longus

Skin wrinkled
by action of
palmaris brevis

Styloid process
of ulna

FIG. 77. Resisting hands showing

Flexion-abduction (commencement of opposition) of carpo-metacarpal joint of right thumb against resisting finger

Abduction of left forefinger against resistance

Extension-abduction of carpo-metacarpal joint of left thumb against resisting little finger, also extension of the other thumb joints

Edge of dorsal expansion of extensor tendon in attempted flexion by dorsal interosseous and lumbrical at metacarpophalangeal joint, while interphalangeal joints are fixed as shown in left middle finger

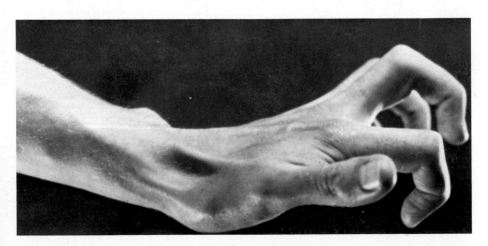

FIG. 78. The hollow between extensor pollicis brevis and longus is termed the 'anatomical snuff-box'

FIG. 79

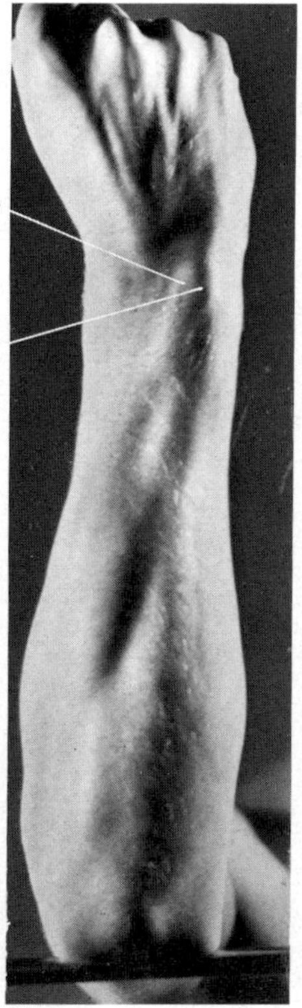

FIG. 80

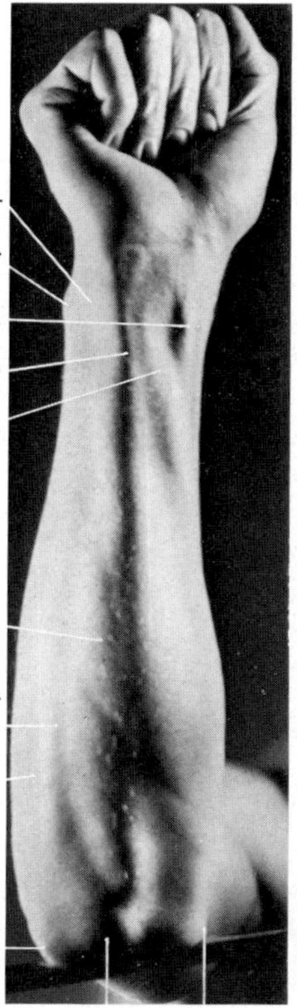

FIG. 81

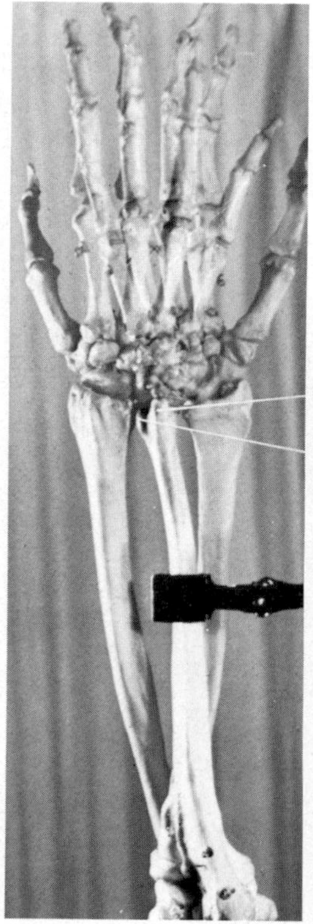

Tendon of extensor
carpi ulnaris

Styloid process
of ulna

Site of styloid pro-
cess of ulna

Lateral aspect of
head of ulna

Flexor carpi radialis

Flexor carpi ulnaris
tendon

Flexor digitorum
sublimis

Flexor carpi ulnaris
belly

Ulna's dorsal border
subcutaneous in
this furrow

Extensor carpi
ulnaris

Latera epicondyle

Olecranon   Medial
epicondyle

Styloid process
of ulna

Lateral aspect of
head of ulna

The ulna unlike the radius does not rotate markedly in pronation and supination of the forearm

The position of the lower end of the ulna remains the same (practically) in supination and pronation of the forearm
In Fig. 79 the tendon of extensor carpi ulnaris lies in the groove between the styloid process and the head of the ulna, which is not visible or palpable in full supination.
In Fig. 80 note the site of the styloid process of ulna which is not visible but palpable in full pronation
One of the student's difficulties in becoming convinced that the head and styloid process of the ulna do not rotate is due to the movement of the skin and tendons during pronation and supination of the forearm

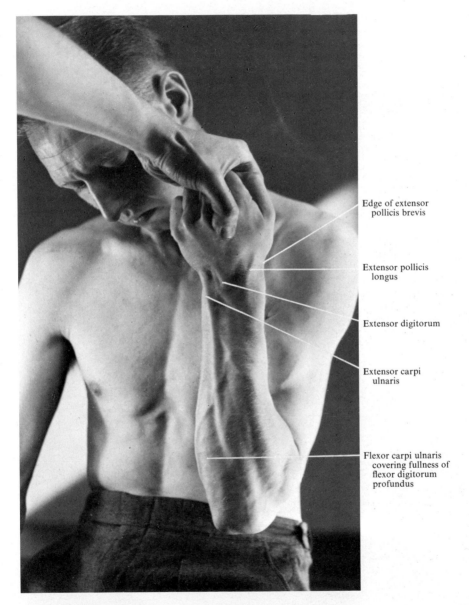

Edge of extensor
pollicis brevis

Extensor pollicis
longus

Extensor digitorum

Extensor carpi
ulnaris

Flexor carpi ulnaris
covering fullness of
flexor digitorum
profundus

FIG. 82. Extension of wrist against resistance

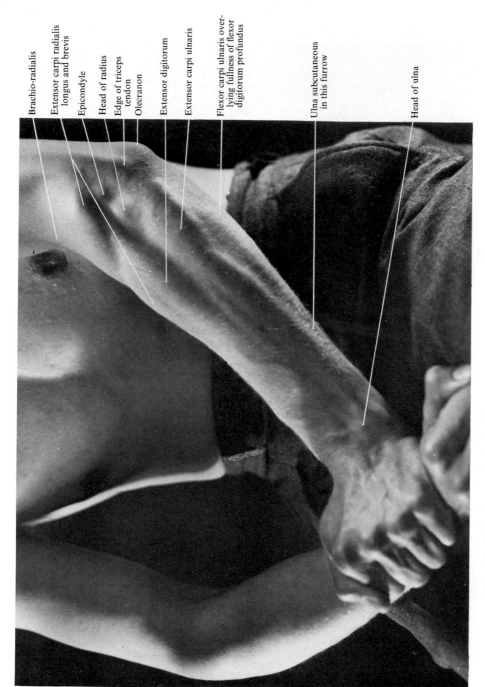

Brachio-radialis

Extensor carpi radialis longus and brevis

Epicondyle

Head of radius

Edge of triceps tendon

Olecranon

Extensor digitorum

Extensor carpi ulnaris

Flexor carpi ulnaris over-lying fullness of flexor digitorum profundus

Ulna subcutaneous in this furrow

Head of ulna

FIG. 83. Relations at lateral side of elbow with forearm pronated

FIG. 84. Valves demonstrated in the veins by thumb arresting venous return flow

Why not try this simple experiment for yourself?

46

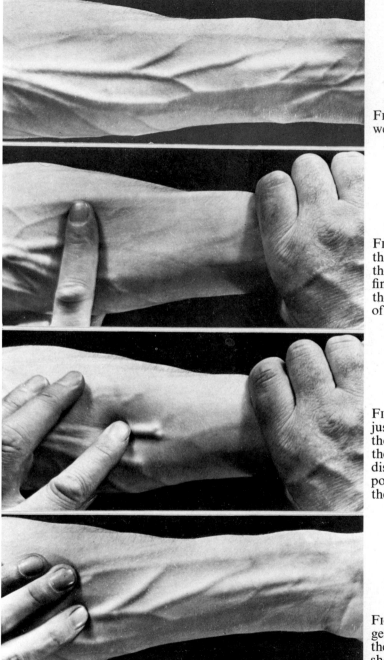

FIG. 85. The veins show
well when the arm has
been dependent

FIG. 86. The hand stops
the venous return up
the forearm while the
finger has just swept
the blood upwards out
of the veins, leaving
them empty

FIG. 87. The finger has
just been lifted and
then swept down along
the vein in order to
distend the valve
pockets which prevent
the blood descending
further

FIG. 88. While the fin-
gers are kept in position
the hand is removed to
show how the blood as-
cends through the veins

Demonstration of venous valves and of the course of the blood through the veins
(From Harvey)

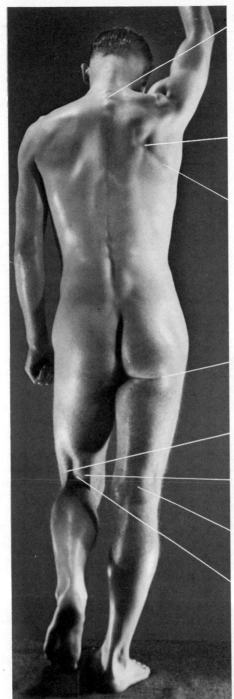

Prominence due to spine
of first thoracic vertebra

Triangular area at medial
end of spine of scapula

Line of vertebral border
of scapula

Fold of buttock

Biceps

Semitendinosus

Popliteal swelling

Popliteal fossa

FIG. 89

Note the positions of both scapulae, the disappearance
of the buttock fold when the limb is flexed, the formation
of a popliteal fossa when the knee is bent, and its
replacement by a swelling when the knee is straight

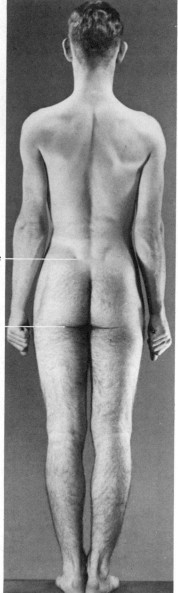

Dimple indicates site
of posterior superior
iliac spine

Fold of buttock

FIG. 90
Contrast with Fig. 91

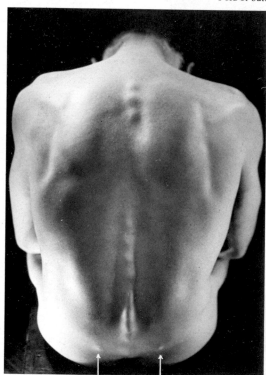

Prominences caused by posterior superior iliac spines
FIG. 91

When the back is bent forward the position of the posterior
superior iliac spines is shown by prominences instead of the
dimples observed when the subject is erect, as in Fig. 90

ASSOCIATED ACTION OF SPINAL MUSCLES IN MOVEMENTS OF THE LOWER LIMB

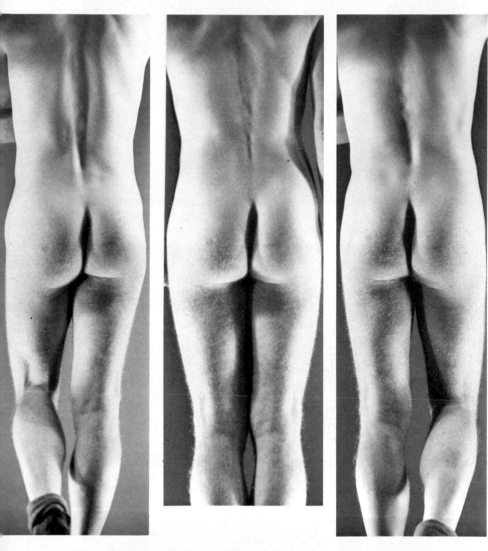

FIG. 92. Standing on right   FIG. 93. Standing on both   FIG. 94. Standing on left
foot   feet   foot

The spinal muscles in the loins contract on the side of the raised foot. Compare Figs. 92, 93, 94
The alternate contraction and relaxation, first one side then the other, is easily felt by placing the hands on one's back during walking

FIG. 97. Activity of serratus anterior, latissimus dorsi and rectus abdominis against resistance

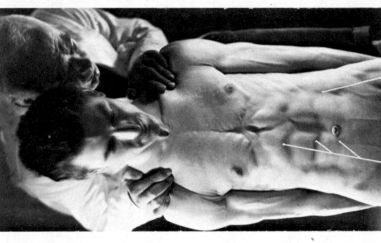

Tendinous intersections of rectus abdominis   Linea semilunaris

FIG. 96. Rectus abdominis flexing the trunk against resistance

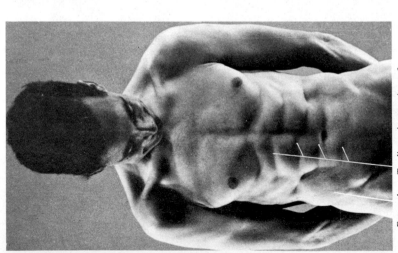

External oblique   Tendinous intersections of rectus abdominis

FIG. 95. Muscles of abdominal wall 'set'

Note the fullness of external oblique bulging over the groove of the groin

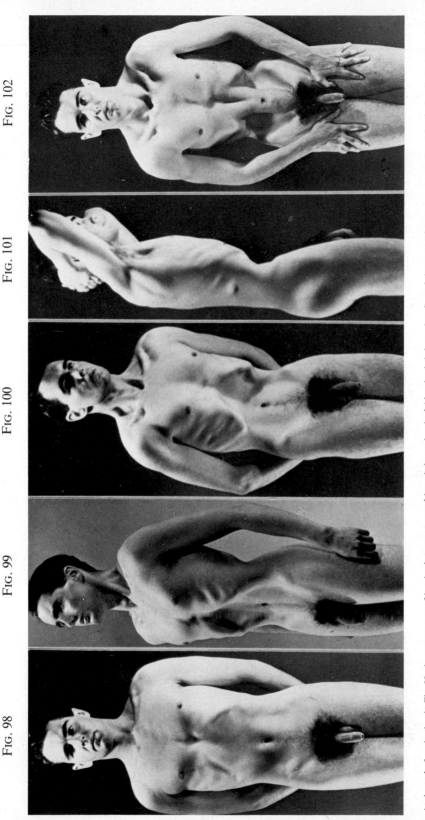

FIG. 98   FIG. 99   FIG. 100   FIG. 101   FIG. 102

At the end of expiration in Fig. 98, the movements of inspiration are executed but air is not inspired (the glottis is kept closed), and the atmosphere presses upon the abdominal wall; the intercostal spaces, the supra-sternal and the supra- and infra-clavicular fossae, Figs. 99, 100, 101. In this condition the subject, by flexing the trunk against the resistance of the arms on the thighs, causes the rectus abdominis muscles to stand out against the collapsed abdomen, Fig. 102.

52

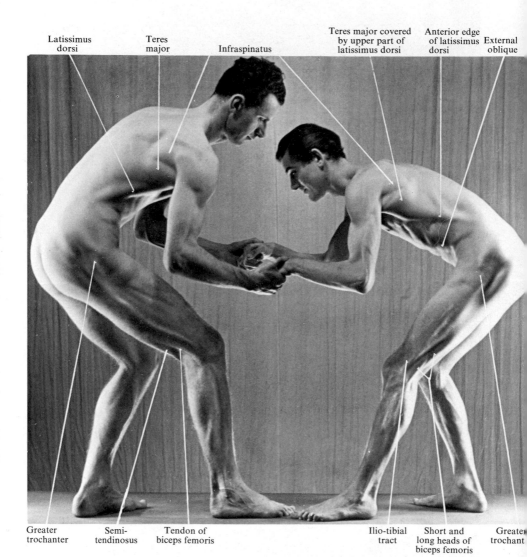

Latissimus dorsi   Teres major   Infraspinatus   Teres major covered by upper part of latissimus dorsi   Anterior edge of latissimus dorsi   External oblique

Greater trochanter   Semi-tendinosus   Tendon of biceps femoris   Ilio-tibial tract   Short and long heads of biceps femoris   Greater trochant

FIG. 103. Activity of latissimus dorsi in both subjects pulling upper arm to the trun (extension of shoulder joint) against resistance

FIG. 104. Different view of the models in Fig. 103, unlabelled for comparison

54

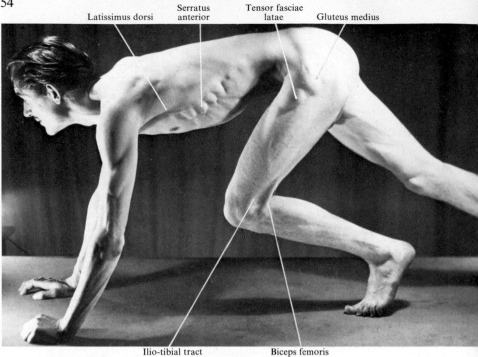

Latissimus dorsi   Serratus anterior   Tensor fasciae latae   Gluteus medius

Ilio-tibial tract   Biceps femoris

FIG. 105. Activity of serratus anterior preventing scapula being thrust back when arm are supporting the weight of the body

FIG. 106. Activity of serrat anterior similar to that Fig. 105 above

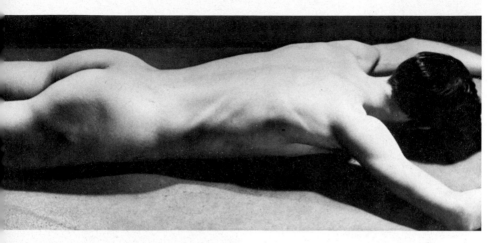

FIG. 107. Subject at rest—post-vertebral and gluteal muscles flaccid

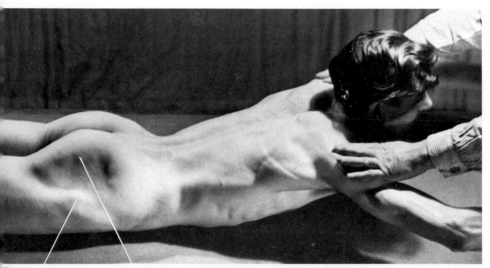

Greater trochanter   Gluteus maximus

FIG. 108. Subject raising arms and trunk from table against resistance (extension of vertebral column and hip joint). Note the activity of the gluteal and post vertebral muscles in contrast with Fig. 107

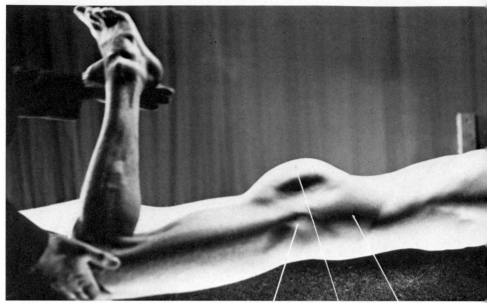

Greater trochanter   Gluteus maximus   Gluteus medius

Fig. 109. Gluteus maximus rotating hip joint laterally against resistance, while gluteus medius is relaxed

Compare Fig. 110

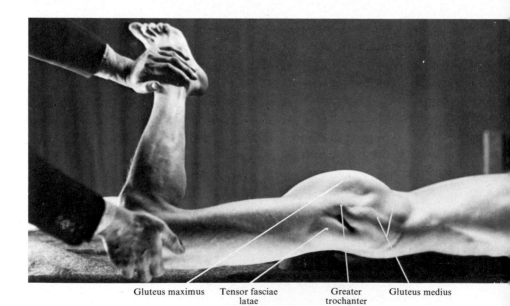

Gluteus maximus   Tensor fasciae   Greater   Gluteus medius
latae   trochanter

Fig. 110. Gluteus medius rotating hip joint medially against resistance, while gluteus maximus is relaxed

Compare Fig. 109

Adductor longus Gracilis   Sartorius   Vastus medialis

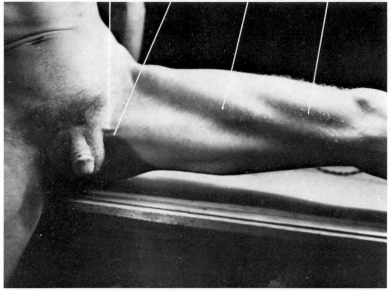

FIG. 111. Femoral triangle outlined by sartorius and adductor longus muscles and inguinal groove

Semitendinosus   Gracilis covering semimembranosus

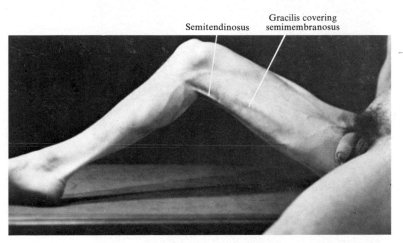

FIG. 112. Flexion of knee against resistance of heel on table. Sartorius, slightly active, tends to sink into relaxed quadriceps femoris

Sartorius   Valve in long saphenous vein   Sartorius

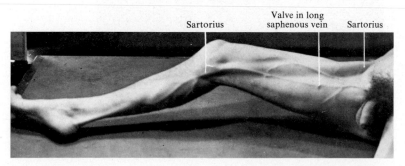

FIG. 113. The oblique course of sartorius across the thigh

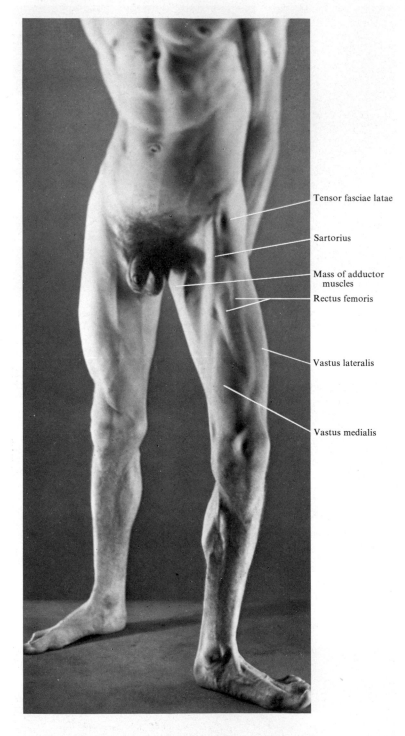

Tensor fasciae latae

Sartorius

Mass of adductor
muscles

Rectus femoris

Vastus lateralis

Vastus medialis

FIG. 114. Subject 'setting' muscles of thigh

See different views in Figs. 115, 116, 117

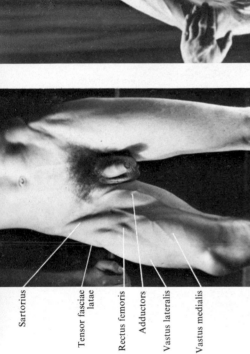

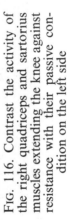

Sartorius

Tensor fasciae latae

Rectus femoris

Adductors

Vastus lateralis

Vastus medialis

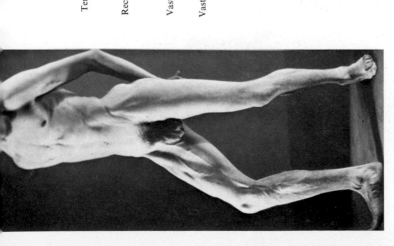

FIG. 115. Muscles of thigh 'set' and adductors pulling against resistance of heel on floor

Note valve pocket in long saphenous vein in upper part of thigh. Compare Figs. 114, 116, 117

FIG. 116. Contrast the activity of the right quadriceps and sartorius muscles extending the knee against resistance with their passive condition on the left side

Compare Figs. 114, 115, 117

FIG. 117. Adduction of thigh against resistance showing the mass of the adductor muscles

Compare Figs. 114, 115, 116

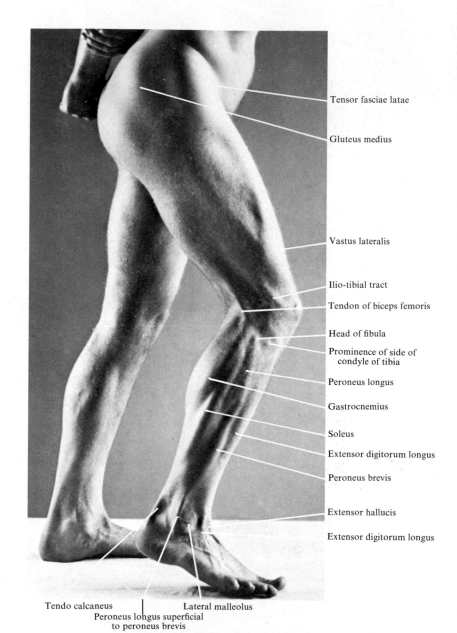

Tensor fasciae latae

Gluteus medius

Vastus lateralis

Ilio-tibial tract

Tendon of biceps femoris

Head of fibula

Prominence of side of
    condyle of tibia

Peroneus longus

Gastrocnemius

Soleus

Extensor digitorum longus

Peroneus brevis

Extensor hallucis

Extensor digitorum longus

Tendo calcaneus   Lateral malleolus
        Peroneus longus superficial
            to peroneus brevis

FIG. 118. Subject 'setting' muscles of leg

Note the concavity of the popliteal fossa behind the flexed knee and the
fullness of this region behind the straight knee

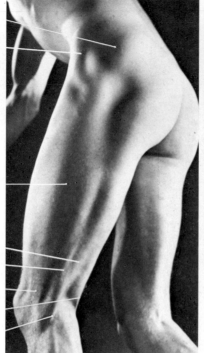

Gluteus medius

Tensor fasciae latae

Vastus lateralis

Biceps femoris { long head / short head

Ilio-tibial tract

Semitendinosus

Biceps femoris tendon

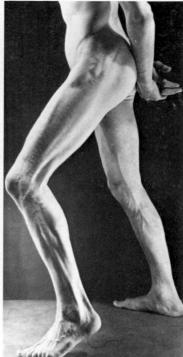

FIG. 119. Lateral and posterior aspect of thigh

FIG. 120. Subject 'setting' muscles of thigh and leg

Unlabelled for comparison with Fig. 118

FIG. 121. Ligamentous action of the hamstring muscles (tight hamstrings) preventing full flexion at the hip joint in the endeavour to touch the toes while the knee is extended

Contrast with Figs. 122 and 123

FIG. 122. Full flexion secured at the hip joint when the knees are flexed

Contrast with Fig. 121

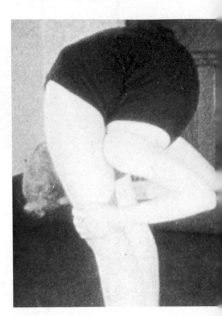

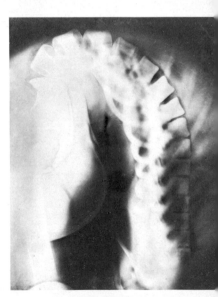

FIG. 123. Remarkable relaxation of the hamstring muscles is required to secure full flexion at the hip joint in this perfect high kick

Contrast with Fig. 121

FIGS. 124 and 125. Relaxation of muscles of anterior abdominal wall in this extreme back bend. There is greater potential for backward than forward bending even in the normal untrained person. Note the vault of the skull between the femora

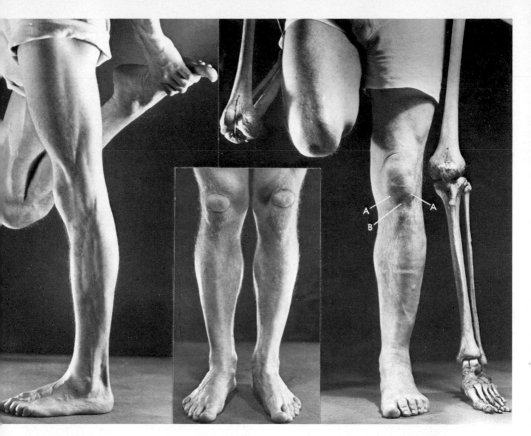

FIG. 126

FIG. 127
Left quadriceps contracted,
right relaxed

FIG. 128
Infrapatellar pad of fat A,
on each side of ligamentum
patellae B

Edges of lateral condyles of femora

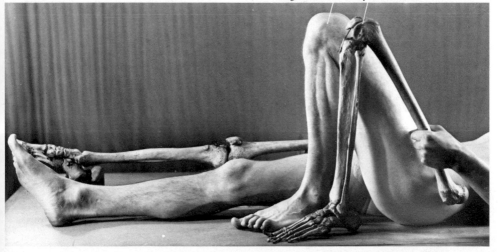

FIG. 129

Figs. 126 to 129 note how the patella recedes between the femoral condyles during flexion of
knee, thereby conferring a smooth rounded contour upon the joint, and advances, thrust
ward by the condyles, during extension. This illustrates its function as a variable gliding
er gaining power towards the end of extension, speed being quicker at the outset of the
vement. The level of the patella differs in the extended knee accordingly as the quadriceps
is contracted or relaxed

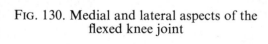

Tendon of   Condyles of tibia   Condyle of
patella   medial   lateral   femur   Head of fibula

FIG. 130. Medial and lateral aspects of the flexed knee joint

Edge of medial ligament

FIG. 131. Medial aspect of knee joint

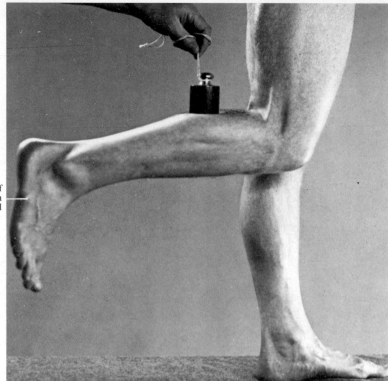

Tubercle of
base of fifth
metatarsal

FIG. 132. A small weight sinks into the calf muscles when the knee is
flexed in the absence of resistance

Compare with Fig. 133

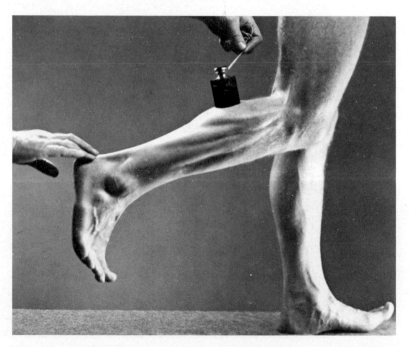

FIG. 133. When the knee is flexed against resistance the calf muscles
are more active and the weight rides upon them

See Fig. 118 to identify muscles

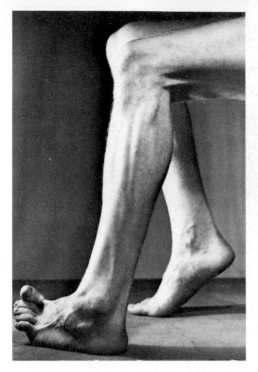

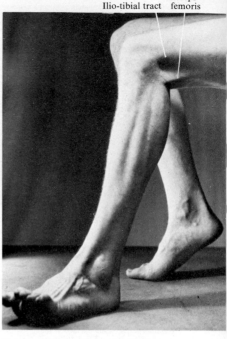

FIG. 135. Lateral rotation of leg upon thigh

Note the increased interval between the tendon of biceps femoris and the ilio-tibial tract covering vastus lateralis

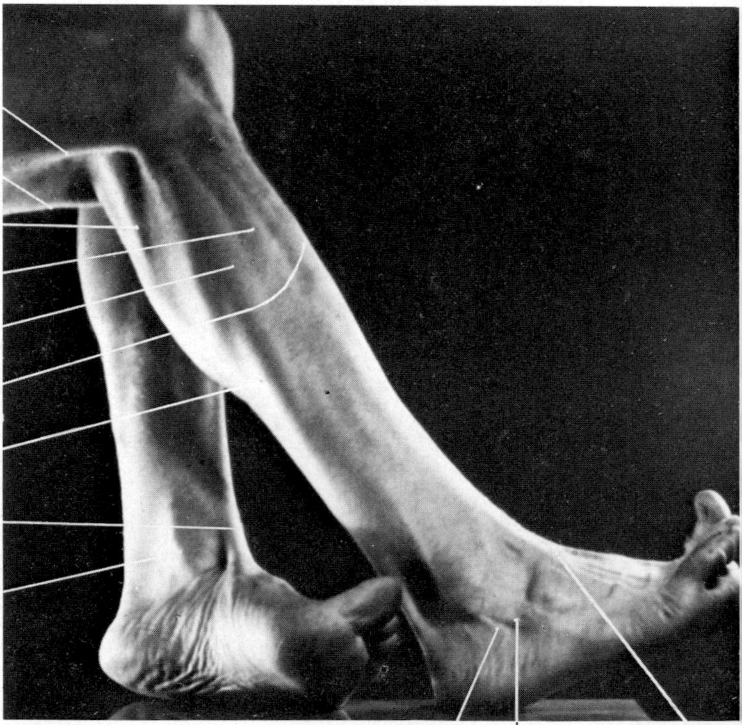

FIG. 134. Medial rotation of leg upon thigh

Contrast the position of the tendon of biceps femoris and ilio-tibial tract with Fig. 135

FIG. 136. Tibialis anterior and posterior inverting left foot while peronei evert right foot and the extensors dorsiflex its toes

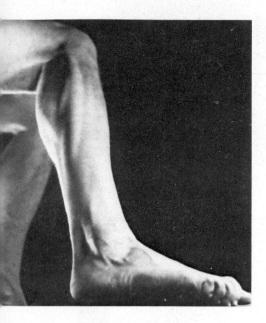

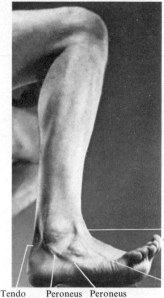

Extensor
digitorum
longus

Extensor
digitorum
brevis

Tendo   Peroneus   Peroneus
calcaneus   longus   brevis

FIG. 137. Attempted eversion with plantar
flexion of foot

Peronei and calf muscles active. Contrast with Fig. 138

FIG. 138. Attempted eversion with
dorsiflexion of foot

Tendons of peronei and extensor digitorum longus
stand out actively. Contrast with Fig. 137

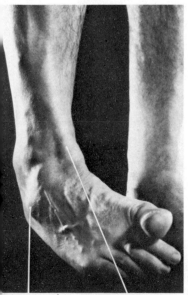

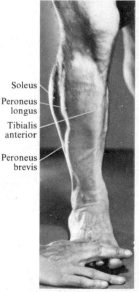

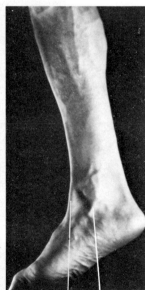

Soleus
Peroneus
longus
Tibialis
anterior
Peroneus
brevis

Calcaneum   Tibialis anterior

Tibialis anterior and posterior

FIG. 139. Inversion of right foot
causing antero-lateral edge of
calcaneum to become prominent

FIG. 140. Dorsiflexion of
foot against resistance

Note large belly of tibialis anterior in
upper part of leg

FIG. 141. Inversion of foot
against resistance

Note that tibialis anterior and
posterior combine to invert the foot
although they oppose each other in
dorsi- and plantar flexion of the foot

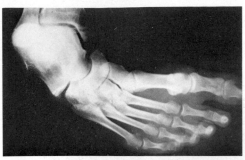

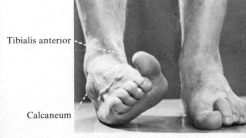

Tibialis anterior

Calcaneum

FIGS. 142 and 143. Radiograph and photograph of inverted foot showing prominence caused by antero-lateral edge of calcaneum. See also Fig. 139

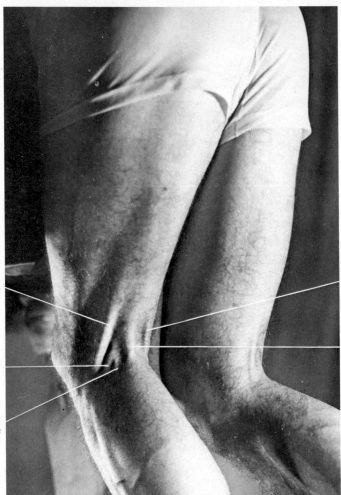

Biceps femoris

Semitendinosus on semi-membranosus

Medial popliteal nerve

Lateral popliteal nerve

Common trunk of sural communicating nerve and lateral cutaneous nerve of calf

FIG. 144. Superficial relations in the popliteal fossa

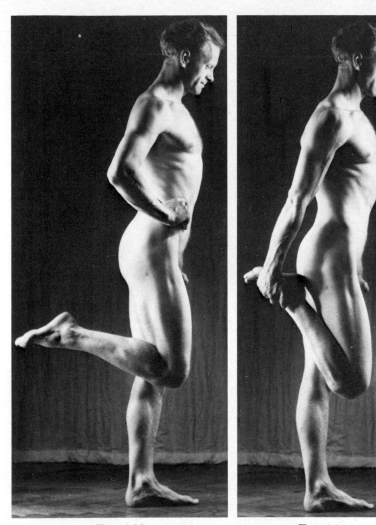

FIG. 145   FIG. 146

MUSCLES OF SHORT OR INSUFFICIENT ACTION

The fibres of biceps femoris, semimembranosus, semitendinosus and
gastrocnemius are too short to contract sufficiently to effect full flexion
of the knee joint; the term active insufficiency is applied to this state of
the muscles. The hamstring muscles also exhibit passive insufficiency or
ligamentous action when they are unable to extend sufficiently to allow
full flexion of the hip joint with the knee extended. Contrast Figs.
121 and 123

FIG. 147. The extent of the female breast—from the mid-sternal to the mid-axillary line, and from the second to the seventh rib

Arms being adducted against resistance to show relation of mammary gland to axillary folds. Note that the breast rests upon pectoralis major, curves round it laterally on to serratus anterior, extends into the axilla (axillary tail) and descends on to external oblique and rectus abdominis

 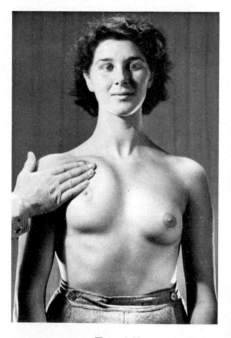

FIG. 148   FIG. 149

Note the alteration in the level of the nipple and breast caused by raising the arm, Fig. 148, and by traction upon the skin Fig. 149. This mobility is also well marked in the male

Fig. 150. Virgin breast of woman 18 years of age

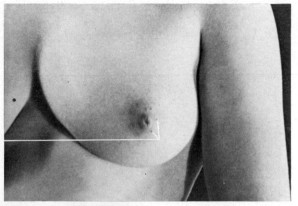

Areolar glands

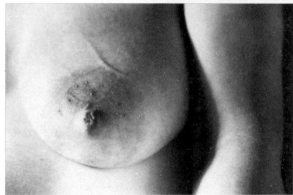

Fig. 151. Lactating breast of mother (with baby seven days old); larger and more heavily-pigmented areola

Areolar glands

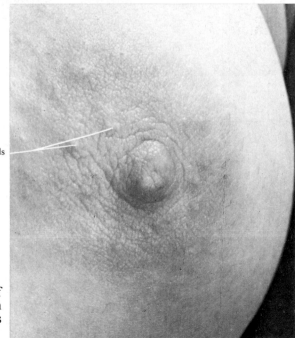

Fig. 152. Enlarged view of nipple and areola, in virgin breast of woman 20 years of age

FIG. 153

ROUGH CONTOUR IN THE MALE

A thinner layer of subcutaneous fat, heavier musculature and larger and rougher bones stamp a harder, angular impression upon the frame in the male; contrast Figs. 153 and 154 with Figs. 155 and 156

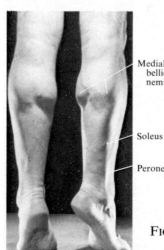

Medial and lateral
bellies of gastroc-
nemius

Soleus

Peronei

FIG. 154. Rising upon the toes by the action of the calf muscles in the male subject

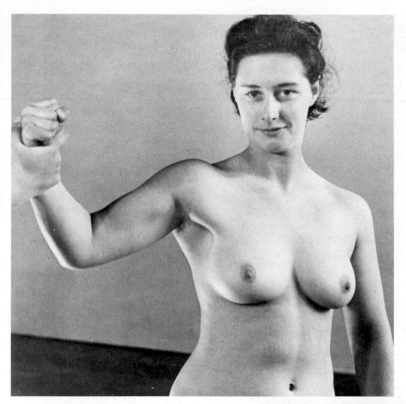

FIG. 155

ROUNDED CONTOUR IN THE FEMALE

A much thicker layer of adipose tissue and smaller, smoother bones confer soft curves upon the form in the female, in whom even powerfully-developed muscles are not usually markedly delineated subcutaneously; contrast Figs. 155 and 156 with Figs. 153 and 154

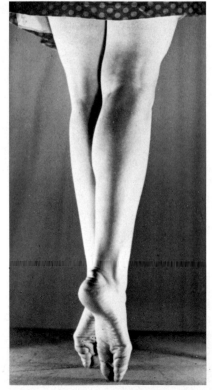

FIG. 156. Ballet dancer on the 'points'

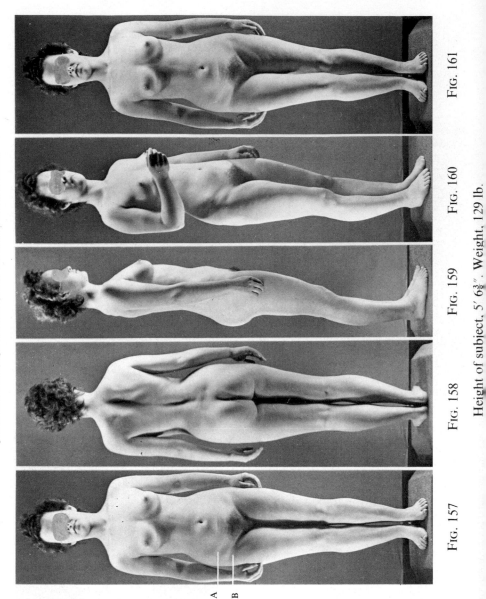

Note the skin crease, Holden's line B, diverges from the groove of the groin A, and runs laterally to the depression over the greater trochanter (Figs. 157, 160, 161, the last with flexed hip and knee)

FIG. 157   FIG. 158   FIG. 159   FIG. 160   FIG. 161

Height of subject, 5' 6¾". Weight, 129 lb.

COMPARISON OF MALE AND FE-MALE FORMS (Figs. 147 to 177)

The frame in woman is more rounded and has more fat, especially in buttocks, flanks, breasts, hips and thighs—fatty tissue 28% in woman, 18% in man; muscles 36% in woman, 42% in man.

She has a comparatively long body and short limbs and a lower centre of gravity.

Her lumbar lordosis is more pronounced.

Her thoracic dimensions are less, and therefore her lung capacity less.

She has a comparatively large abdominal surface.

Weaker shoulders, less than (or the same as) hip width, contrast in man, with stronger shoulders, wider than the hips. Her arm is more cylindrical, his, flatter from side to side, and her forearm is flatter antero-posteriorly.

The one external dimension actually greater in woman than in man is the circumference of the upper part of the thigh, due to the thicker fat deposit occur-ring mainly on the lateral aspect (although the adductor muscles are also well developed); this fat deposit causes the greatest hip width to lie inferior to the level of the greater trochanter, while in the male the greatest hip width lies at the level of the greater trochanter. The thigh in woman is shorter, and as the knee circumference is almost

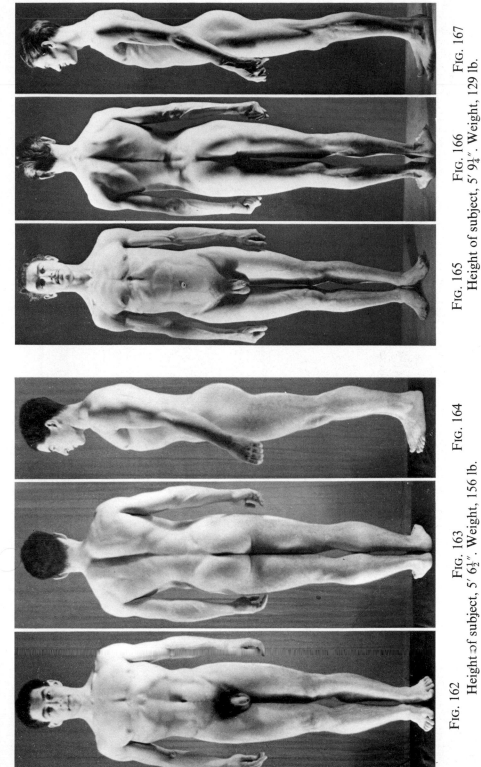

into the fat of the flank, so that the buttocks may appear to ascend to the waist line

FIG. 162   FIG. 163   FIG. 164
Height of subject, 5′ 6½″. Weight, 156 lb.

FIG. 165   FIG. 166   FIG. 167
Height of subject, 5′ 9¼″. Weight, 129 lb.

Height of subject, 5′ 3″. Weight, 133 lb.

Groove of groin

Skin crease

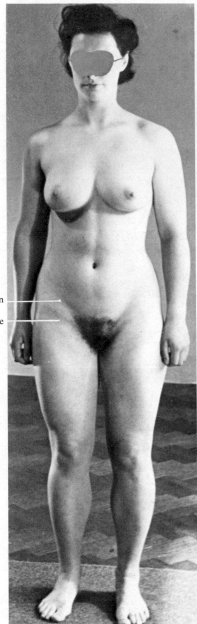 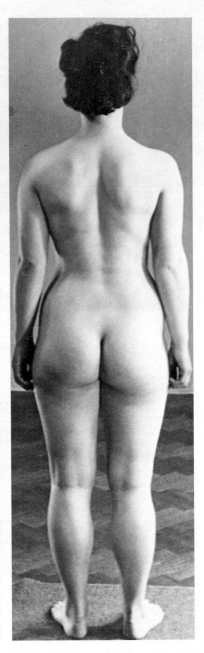

FIGS. 168 and 169. Proportions of shoulders and hips typically feminine

Contrast with Figs. 170 and 171

See text on pages 74 and 75

Height of subject, 5′ 5½″. Weight, 150 lb.

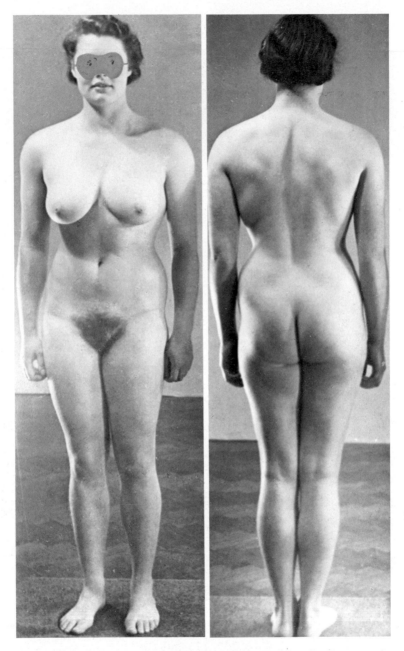

FIGS. 170 and 171. Variation with greater shoulder development in
excellent swimmer

Contrast with Figs. 168 and 169

See text on pages 74 and 75

Height of subject, 5′ 7″. Weight, 140 lb. Age 20 years

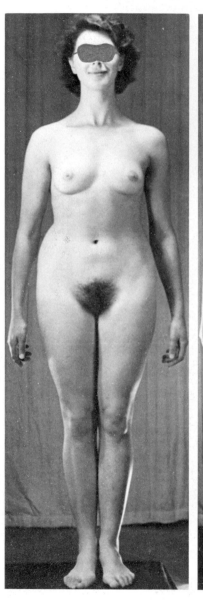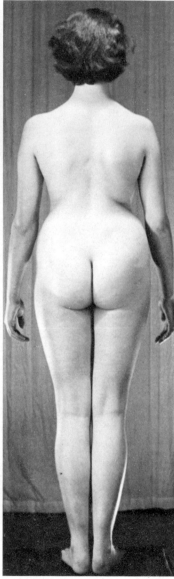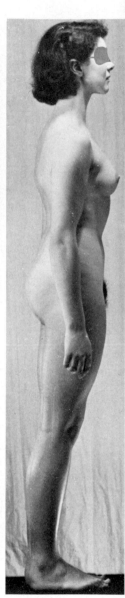

FIGS. 172, 173 and 174

Contrast with Figs. on pages 74, 76, 77 and 79

See text on pages 74 and 75

Height of subject, 5′ 1″. Weight 124 lb. Age 20 years

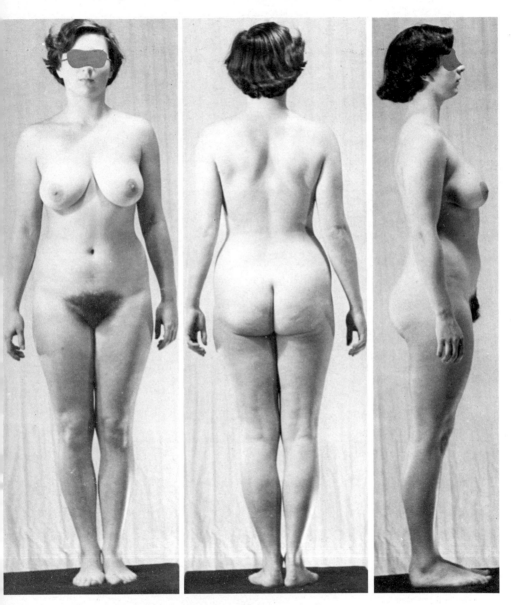

FIGS. 175, 176 and 177

Contrast with Figs. on pages 74, 76, 77 and 78

See text on pages 74 and 75

FIG. 178

Usually, in distant regard, no white is seen between the upper and lower peripheral edge of the iris and the eyelids, but here the lids are opened wide against the photographic lights—hence the staring effect. Contrast with Fig. 179

Lacrimal papilla

Lacrimal caruncle

Lacrimal papilla

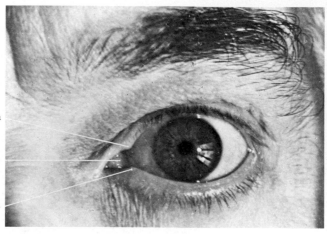

FIG. 179

The palpebral fissure of the eyelids is narrowed against the photographic lights by the encircling orbicularis muscle, which produces the characteristic wrinkling of the skin of the eyelids. Contrast with the opposite effect in Fig. 178. The eyelids meeting at the medial angle of the fissure cover the lacrimal papillae and the triangular interval containing the lacus lacrimalis, seen in Fig. 178

Tarsus of upper eyelid

FIG. 180

Due to the presence of a plate of dense tissue in each lid termed the tarsus, the eyelid may be easily everted to remove specks of grit from the eye. Note the punctum lacrimale, the orifice of the tear duct (lacrimal canaliculus) on the lacrimal papilla in each eyelid

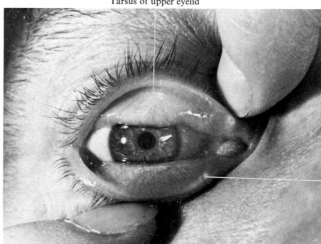

Pun
tum
lac-
rim

FIG. 181

The subject may keep the lid everted, especially the upper one, without assistance

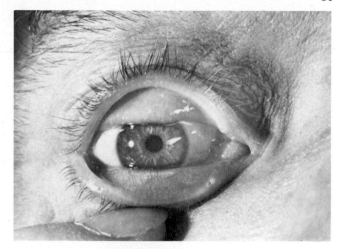

Tarsal glands

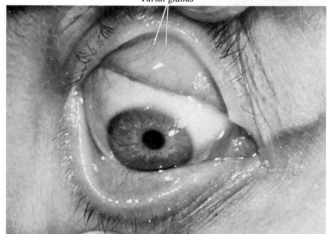

FIG. 182

Eyelids everted and eyeball looking down. Note tarsal glands in parallel rows streaming to the free edge of the eyelids

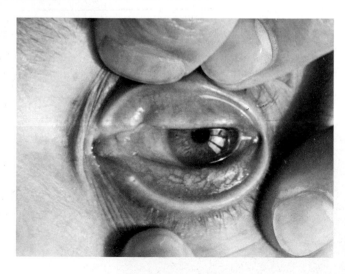

FIG. 183

Complete eversion of both lids

Fig. 184. Under surface of tongue in male 58 years of age

Fimbriated fold

Mass of genioglossus muscle

Tributaries of

profunda vein

Frenulum

Sublingual papillae

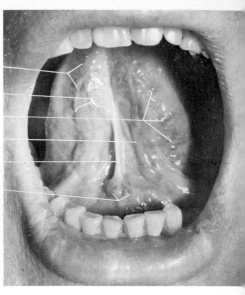

Fig. 185. Same person as in Fig. 184

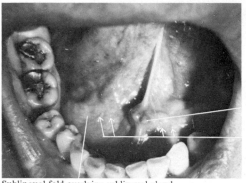

Orifice of submandibular duct on sublingua papilla

Some of the orifices of the ducts of the sublingual gland

Sublingual fold overlying sublingual gland

Profunda vein on the column formed by the genioglossus muscle

Fig. 186. Under surface of tongue in adult female

Sublingual fold overlying sublingual gland

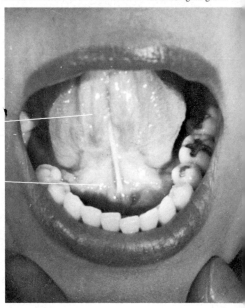

Fig. 187. Arches of oropharyngeal isthmus (pillars of fauces) in male of 58 years, contrast with same person in Fig. 188

Intratonsillar cleft   Semilunar fold

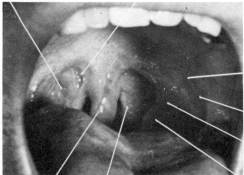

Projection of hamulus of medial pterygoid plate easily felt, often seen

Line of pterygomandibular ligament

Palatoglossal arch

Uvula   Palatopharyngeal arch   Tonsil

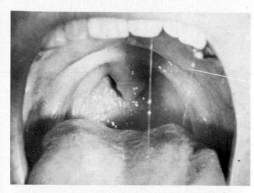

FIG. 188. Note the continuity of the palatoglossal arches from side to side across the base of the uvula (contrast the arches with the same person in Fig. 187)

Projection of hamulus of medial pterygoid plate

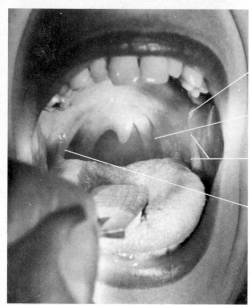

FIG. 189. Arches of oropharyngeal isthmus in adult female

Projection of hamulus of medial pterygoid plate

Tonsil between palatoglossal and palatopharyngeal arches

Line of pterygomandibular ligament

Note the palatoglossus muscle in its arch entering the tongue

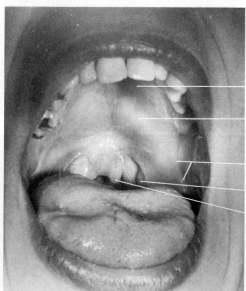

FIG. 190. Note the alteration of the colour of the mucous membrane marking the hard from the soft palate. Same person as in Fig. 189

Hard palate

Soft palate

Line of pterygómandibular ligament

Tonsil between the arches

Uvula

84

FIG. 191.   6 yrs.   FIG. 192.   13 yrs.   FIG. 193.   16 yrs.   FIG. 194.   18 yrs.

3' 9"   5' 3½"   4' 10½"   5' 5"   5' 6½"   5' 5½"   5' 10"

FIGS. 191–194. Twins contrasting growth rate of male and female. Level in height up to 6 years and even up to 10 years, the girl then steals a march upon her brother by 13 years of age, but by 16 years when she has little further to grow, he comes abreast and is well ahead at 18 years. Individuals vary but there is little growth in height after the limb bones cease growing at 20 years though some people may not reach their full height until 25 years.

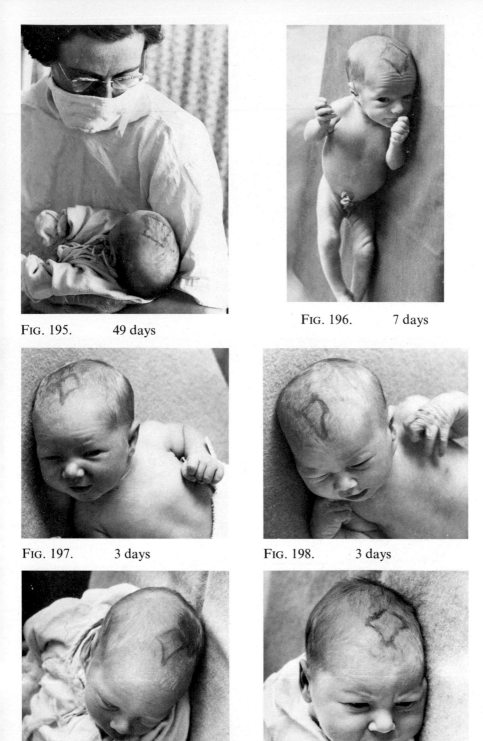

FIG. 195.   49 days

FIG. 196.   7 days

FIG. 197.   3 days

FIG. 198.   3 days

FIG. 199.   23 days

FIG. 200.   8 days

FIGS. 195–200. The diamond shapes outline the anterior fontanelle, a membrane filled gap between the frontal and parietal bones at the bregma, the junction of the coronal and sagittal sutures. Easily felt and often seen, this fontanelle at birth varies markedly in size, from $2\frac{1}{4}'' \times 2\frac{1}{4}''$ to $\frac{3}{4}'' \times \frac{3}{4}''$, and even much less. It is filled by the growth of the bones, usually during the second year, but closure may be delayed in conditions such as rickets and hydrocephalus. Indeed, its clinical importance is to be further stressed—intravenous injection may be made through the fontanelle into the superior sagittal sinus if entry is difficult into a superficial limb vein; pulsation may be felt at the fontanelle which becomes bulging and tense with the raised intracranial pressure of meningitis and hydrocephalus, but depressed and soft with the lowered intracranial pressure of wasting diseases; palpation of its position upon vaginal examination may decide the type of presentation at childbirth

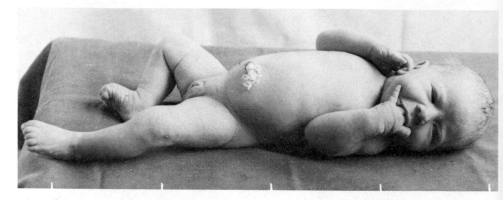

FIG. 201. Note the proportion of the head to the height in this baby of 20 inches and 1 day old. Contrast with Figs. 191 to 194

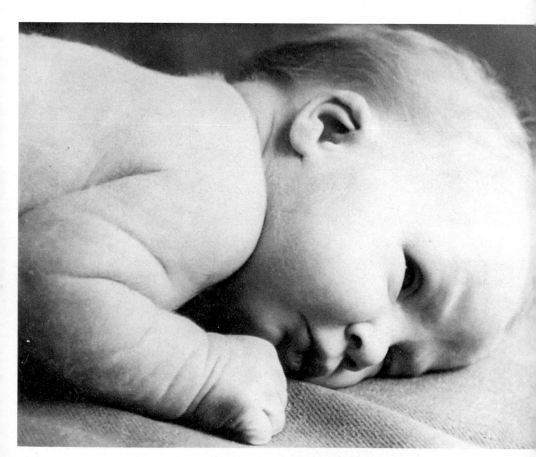

FIG. 202. The fine downy hairs or lanugo, well developed by the seventh month of intr uterine life, are seen here at one day old, on face, trunk and arm; they are more numero in man than in apes, and are shed soon after birth

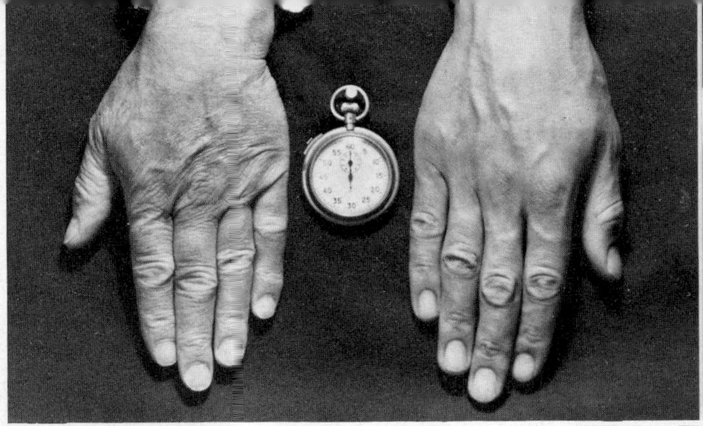

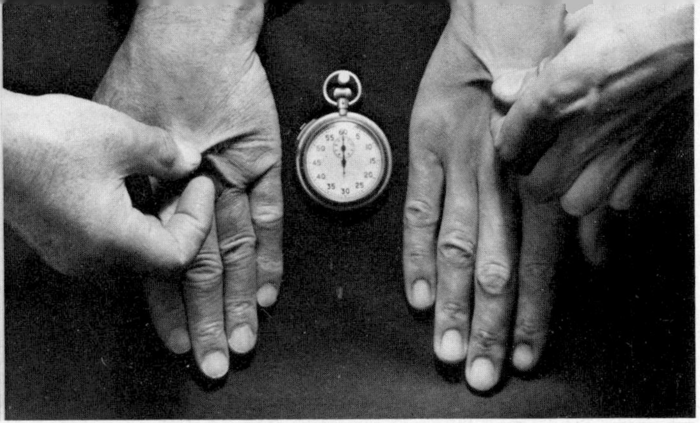

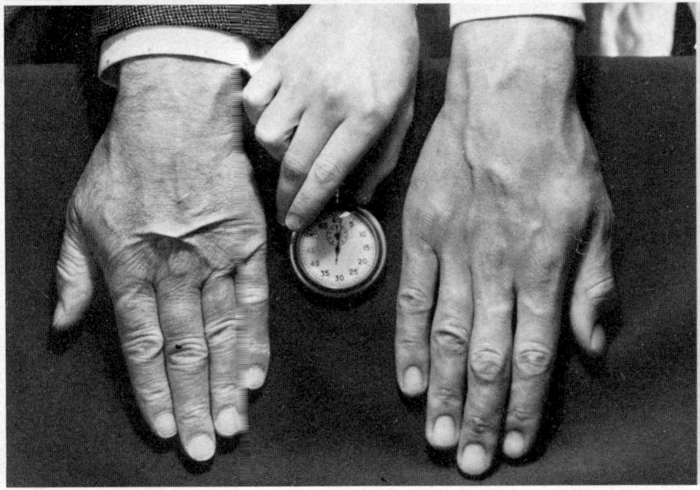

FIG. 206

FIG. 205

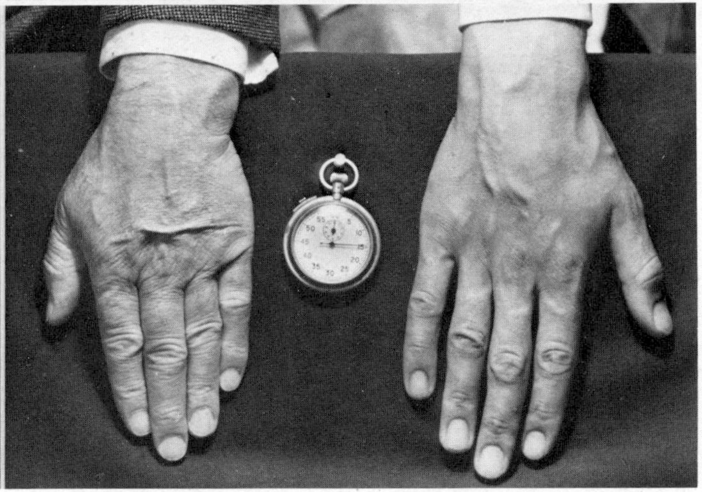

Elasticity of the Skin Contrasted in man of 68 years (Left) and man of 26 years (Right). Fig. 203 at rest; Fig. 204 stretched; Fig. 205 released—returns to normal instantly in man of 26 years; Fig. 206 remains wrinkled 15 seconds later in man of 68 years. It is generally accepted that the skin tends to lose its elasticity with advancing years but there is great variability. See also Figs. 213 to 217

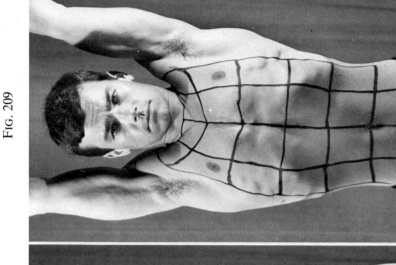

FIG. 207

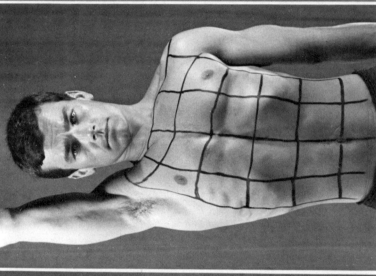

FIG. 208

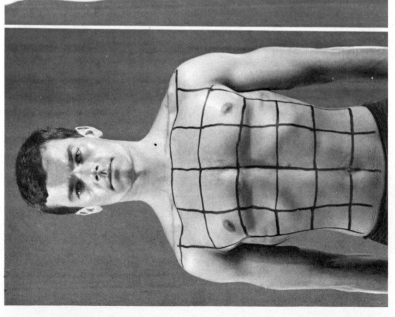

FIG. 209

FIGS. 207–212. Elasticity of the Skin

In Figs. 207 and 210 the trunk is marked by 3″ squares

With the right arm raised, Figs. 208 and 211, the obvious stretching of the skin may be further appreciated by horizontal lines drawn from the ends of the transverse lines on the left side

FIG. 212   FIG. 211 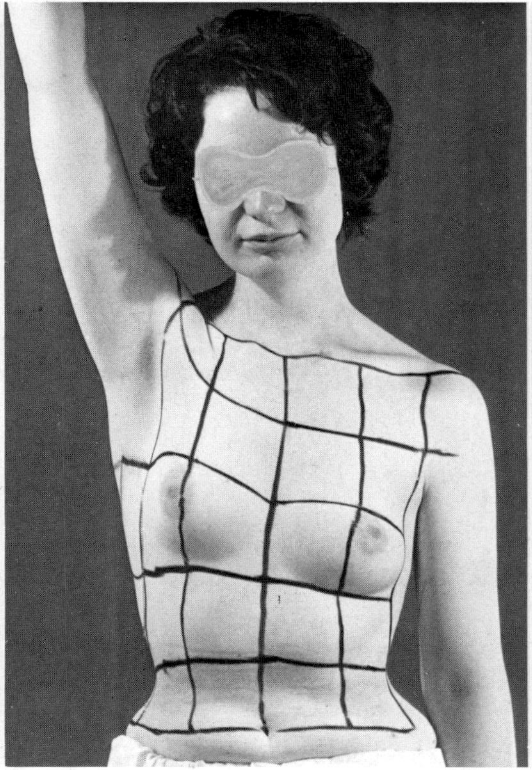   FIG. 210

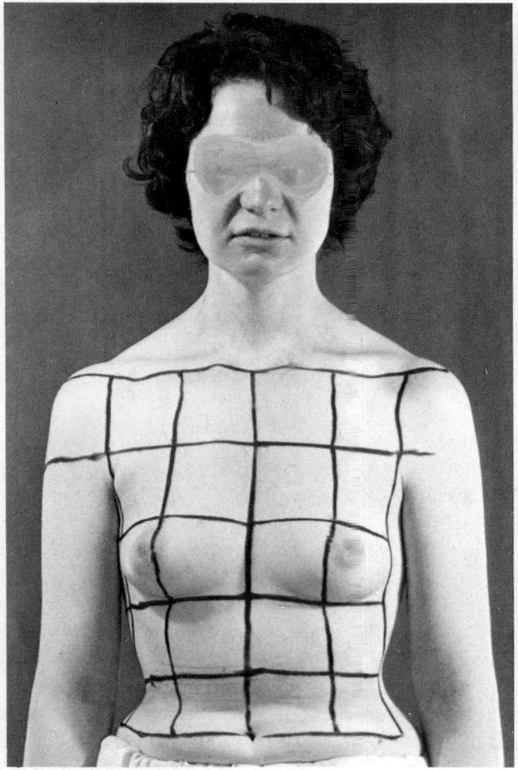

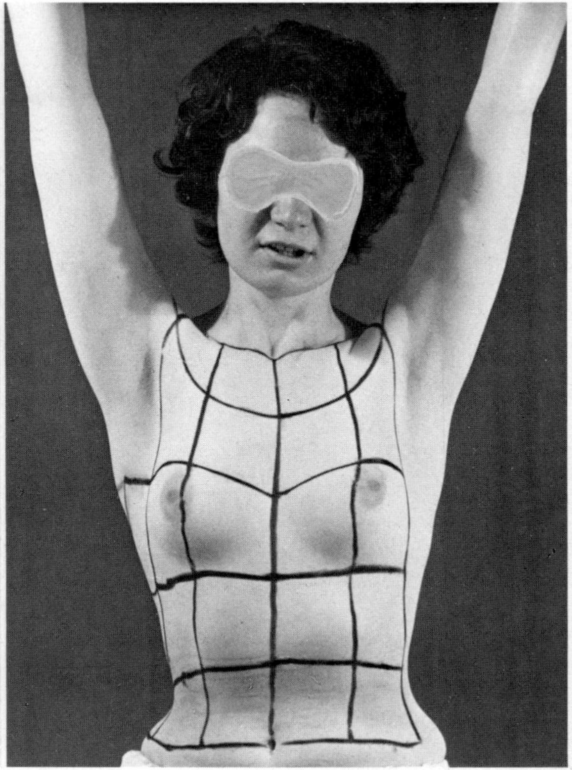

FIGS. 207–212. Elasticity of the Skin
See caption to opposite page

FIGS. 213–217. TIME MARCHES ON

FIG. 213. Child 11 months, Mother 26 years. Note the chubby fat cheeks of the in[...] due to the buccal pad of fa[...] (sometimes in error termed sucking pad). This disappea[...] with the rest of the baby fa[...] but traces may persist in th[...] adult. Contrast the smooth[...] of the infant's face and the wrinkles already apparent under the Mother's eyes. T[...] size of the face relative to t[...] head is smaller in the child[...] in the Mother, the mandibl[...] maxillae, teeth and sinuses having still to develop muc[...] further. The height of the maxillary sinuses, shown between the black spots, is about 1 cm. in the child an[...] just over 3 cms. in the Mot[...] while the floor of the sinus[...] well above the nasal floor i[...] the child and may be 1 cm. below the nasal floor in the adult (the floors are level by eighth year)

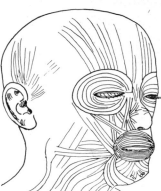

FIG. 214. Direction of the facial muscle fibres

FIG. 215. Age 18. Contrast with Fig. 216

FIG. 216. Age 67: the same person at 18 years is shown in Fig. 215

The expressive facial muscles wrinkle the skin at right ang[...] to their lines of pull, while the advancing years, with attenda[...] loss of fat and of the skin's elasticity, accentuate these wrinkl[...] stamping a characteristic pattern, although it is to be not[...] that the faces of some old people may remain remarkab[...] smooth. Compare the direction of the fibres of the fac[...] muscles in Fig. 214 with the direction of the wrinkles[...] Figs. 216 and 217. See also Figs. 218, 219

FIG. 217. Age 92

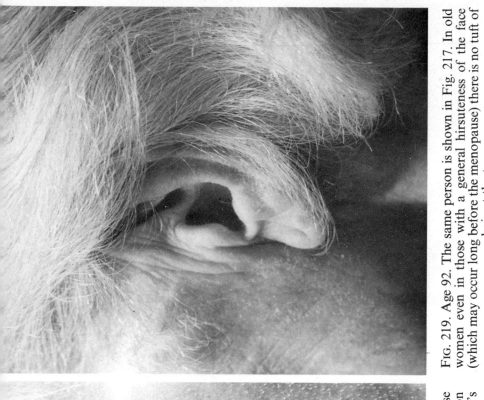

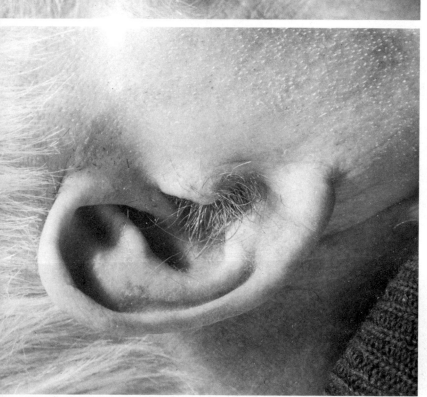

Fig. 218. Age 64. In some men, even in the early thirties, coarse long hairs appear in the nostrils, in the eyebrows and round on to the cheek below the lower eyelid. They also grow in a goat's beard-like tuft from the tragus of the ear (tragus, a goat)

Fig. 219. Age 92. The same person is shown in Fig. 217. In old women even in those with a general hirsuteness of the face (which may occur long before the menopause) there is no tuft of hair at the tragus

FIGS. 220–223. Radiographs and photographs of the foot bearing weight under different conditions

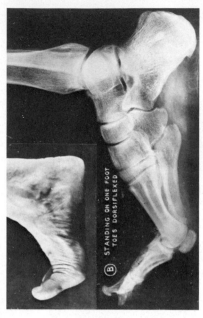

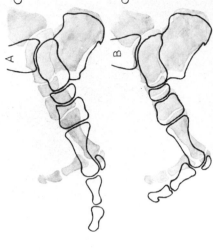

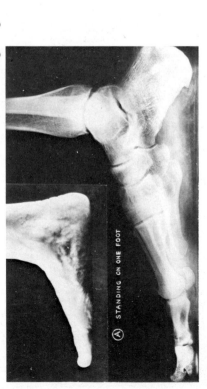

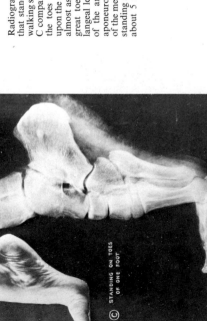

COMPARISON OF A AND C

COMPARISON OF B AND C

Radiographs A and C compared show that standing on the toes as in every walking step, raises the arch, but B and C compared show that dorsiflexion of the toes with the whole weight still upon the rest of the foot raises the arch almost as much, a windlass effect, the great toe extensors pulling the phalangeal lever and shortening the base of the arch by winding the plantar aponeurosis round the drum or head of the metatarsal. The average person, standing at attention, raises his height about 5 mm. merely by dorsiflexing his toes

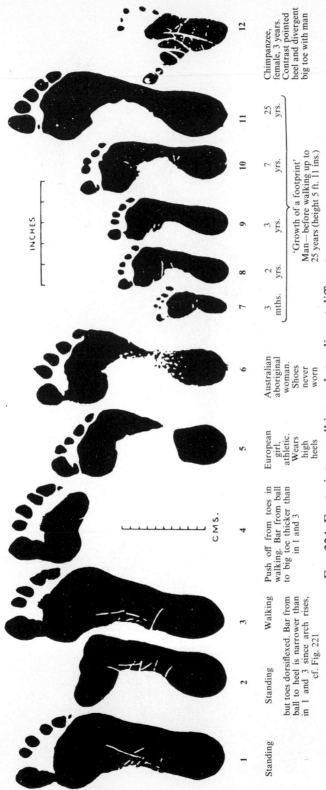

Fɪɢ. 224. Footprints—walking and standing at different ages

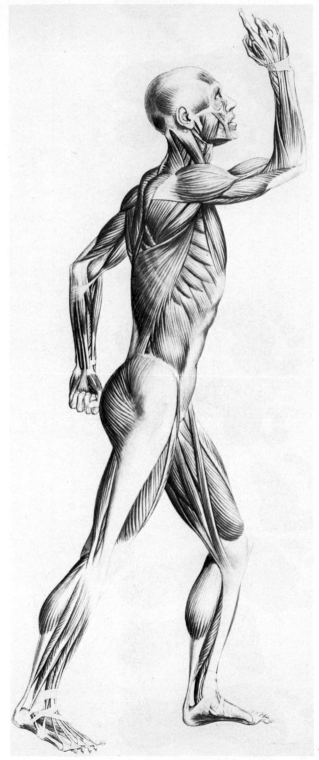

Fig. 225. Surface muscles side view

Figs. 225–227 from *Anatomy of the Human Body*

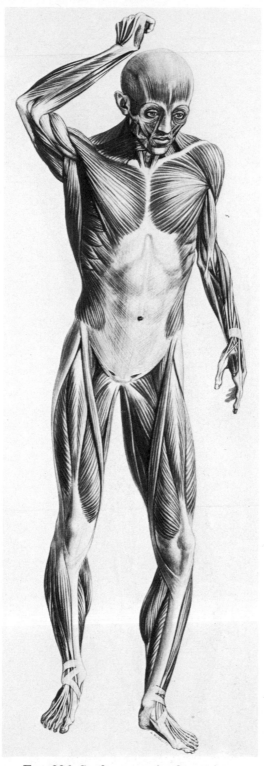

FIG. 226. Surface muscles front view

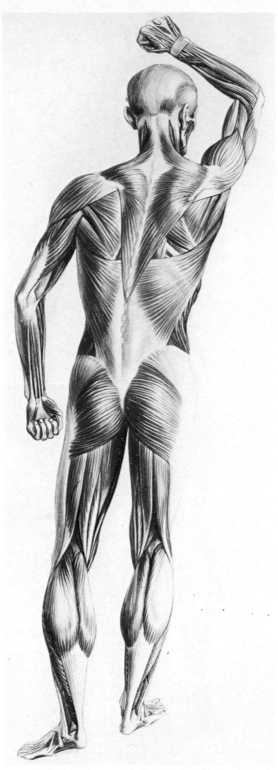

FIG. 227. Surface muscles back view